THE BOOK OF 100 TYPE FACE ALPHABETS

THE BOOK OF 100 TYPE FACE ALPHABETS

A GUIDE TO BETTER LETTERING FOR ARTISTS
AND DESIGNERS OF SIGNS, SHOWCARDS, DISPLAYS, EXHIBITS
AND ADVERTISING LAYOUTS, BY J. I. BIEGELEISEN

Poynter Institute for Media Studies
Library

SEP 16 '86

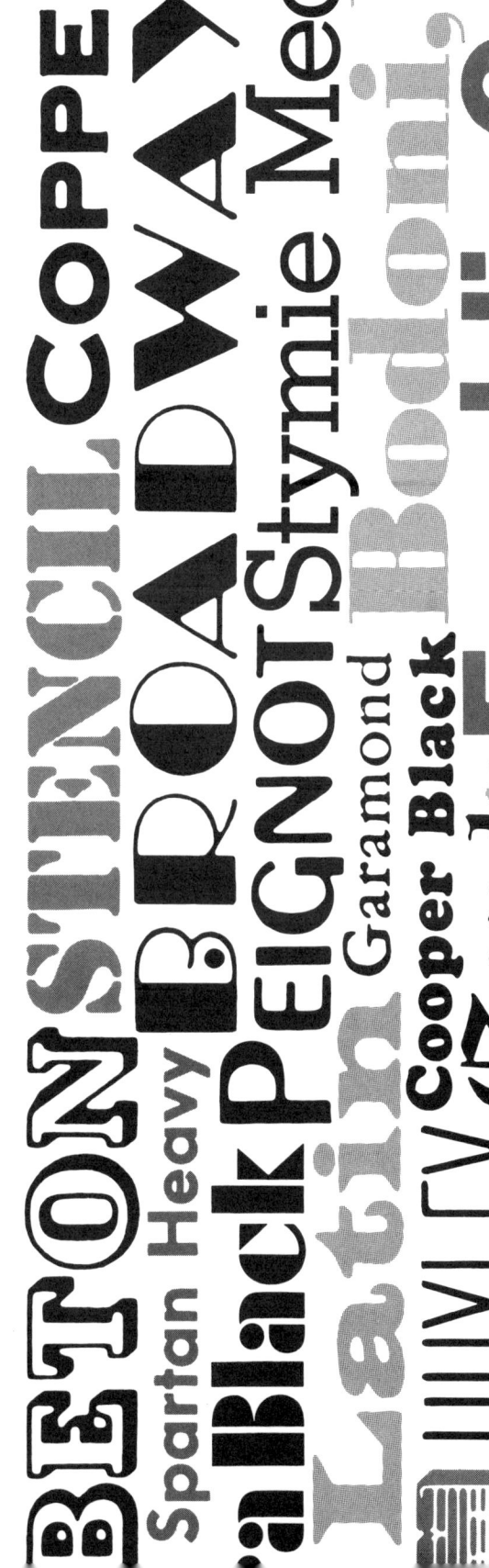

THE SIGNS OF THE TIMES PUBLISHING COMPANY

Copyright: The Signs of the Times Publishing Co. — 1974
Published by The Signs of the Times Publishing Co.
407 Gilbert Avenue, Cincinnati, Ohio 45202, U.S.A.
Library of Congress Catalog Card Number: 65-15691
Printed in Spain, Second printing 1977
International Standard Book Number: 0-911380-03-5

Lovingly dedicated to my wife Esther who served so efficiently and faithfully as my "Girl Friday," every day of the week, for the many, many months it took to get this book ready.

Index and Alphabet Selector

17 **Airport Black**
19 *Allegro*
21 **BALLOON EXTRABOLD**
23 *Bank Script*
25 Baskerville
27 **Bernhard Brushscript**
29 Bernhard Modern Bold
31 *Bernhard Modern Bold Italic*
33 Bernhard Tango
35 **Beton Extra Bold**
37 **BETON OPEN**
39 Bodoni
41 **Bodoni, Ultra**

43 **BROADWAY**
45 *Brush Script*
47 Bulmer
49 CAROLUS
51 Caslon No. 540
53 CENTURY Expanded
55 *Century Expanded Italic*
57 Chisel
59 City Light
61 **Clarendon Bold**
63 *Commercial Script*
65 **Contact Bold Condensed**
67 **Cooper Black**

#	Typeface
69	COPPERPLATE GOTHIC BOLD
71	Corvinus Bold
73	Craw Modern Bold
75	CRISTAL
77	DELPHIAN
79	Dom Casual
81	Egyptian Expanded
83	Engravers Old English
85	Fortune Bold
87	FORUM
89	Franklin Gothic
91	Franklin Gothic Extra Condensed
93	Franklin Gothic Wide
95	Freehand
97	FRY'S ORNAMENTED
99	Futura Black
101	*FUTURA Bold Italic*
103	Futura Display
105	GALLIA
107	Garamond Bold
109	*Garamond Italic*
111	*Gloria*
113	GOLD RUSH
115	Goudy Oldstyle
117	Goudy Text
119	*Goudy*

121	Grotesque No. 9		147	Lucian Bold
123	HEADLINE GOTHIC		149	Lydian
125	Hellenic Wide		151	*Lydian Cursive*
127	HOLLYWOOD		153	Lydian Italic
129	HUXLEY VERTICAL		155	**MICROGRAMMA BOLD**
131	JIM CROW		157	*Murray Hill Bold*
133	Kabel Bold		159	**NEULAND**
135	*Kaufmann Script*		161	News Gothic Bold
137	Latin Wide		163	*Old Gothic Bold Italic*
139	*Legend*		165	Onyx
141	LIBRA		167	PEIGNOT
143	Lilith		169	Perpetua
145	LOMBARDIC INITIAL		171	Playbill

Index and Alphabet Selector

173	PRISMA
175	PROFIL
177	Quillscript
179	Repro Script
181	ROMANTIQUE
183	Rondo Bold
185	Samson
187	SAPPHIRE
189	Spartan Heavy
191	STENCIL
193	Strudimrius
195	Studio
197	Stylescript
199	*Stymie Black Italic*
201	Stymie Medium
203	THORNE
205	Torino
207	Traffon Script
209	TRYLON SHADED
211	Typewriter, Bulletin
213	UMBRA
215	Venus Extrabold Extend'd

Introduction

It is gratifying to both publisher and author of *The Book of 100 Type Face Alphabets* to know that the demand for this book has in no way diminished since its original publication in 1965. Indeed, during the intervening years the special merits of the book have been even more widely acknowledged among professionals and students as one of the basic sources of alphabet references for the workshop, studio and classroom.

The Book of 100 Type Face Alphabets in its present format, has retained the outstanding features of the earlier edition. The alphabets shown are in complete font, in most instances include capital letters, lower case letters and numerals. Each of the alphabets is systematically described in terms of essential characteristics, construction and applications. The distinguishing difference between this edition and the earlier one, is principally in format.

With the ever escalating costs of paper, printing and other related factors involved in the production of a book of this kind, it was found desirable to modify the general format and reduce the page size to effect some measure of economy, thus making it possible to retain the 1965 price. It is important to point out that although the page size is smaller, there has been only a small reduction in the actual size of the alphabet and no change in the type face of the text.. In a sense, the smaller-size book has a built-in favorable feature; it take up less desk space, leaving more elbow room for paints, brushes and accessories.

The "Index and Alphabet Selector", which has been one of the unique and innovative features of the original edition has been retained. It shows, not merely where each type face is located in the book, but gives a sampling of the face for identification and type selection.

I want to take this opportunity to thank ST Publications, the publishers of this book for making the effort in the reprinting of the book to keep the price within reach of not only the commercially successful professionals in the field, but for the young people who are students of lettering and typography and the professionals of the future.

J. I. Biegeleisen
April, 1974

THE BOOK OF 100 TYPE FACE ALPHABETS

ABCDEFGHIJKLMN
OPQRSTUVWXYZ

&!?$1234567890

abcdefghijklmnopqrstuvwxyz

■ ESSENTIAL CHARACTERISTICS

Here is a masculine looking type face that shows its relationship to the Gothic and Futura families. It is in the main a single-thickness letter with slight variations needed for expediency in construction. For instance, the cross bar of the A is thinner than the basic thickness, because there just wouldn't be space enough to keep a uniform thickness throughout. This also applies to a greater or lesser extent to the center cross bars of the E, F, G and H. Because of the massive weight of Airport Black, the counters (inside areas) of the letter forms are rather small compared to the main body of the letter. In the lower case the counters of the round letters are almost perfect circles; the a, you will note, is a typical example. The dots of the i and j are elliptical and placed close to the body of the letter. Also note that the j is nothing more than an elongated i.

■ ESSENTIAL CONSTRUCTION

If this alphabet is to be indicated in rough comprehensive form for layout work it can be easily rendered with a broad carpenter's pencil, well-chiseled and handled flat. A few deft strokes rendered by an expert layout man will be sufficient to unmistakably indicate Airport Black on the working layout. If a paint rendering is required—either for a color comp or semi-finished art, a well-palleted short-haired showcard lettering brush is recommended, ranging in size from No. 4 to No. 12, depending on the desired size of the letter form and the degree of finish needed. For extra finish jobs in color (where photo lettering could not be used) this type face can be meticulously, though not laboriously, rendered with ruling pen, India ink, and white touch-up wherever needed.

■ ESSENTIAL APPLICATIONS

This type face is certainly one of the bread and butter alphabets that should be in the repertoire of every lettering man and layout artist. Airport Black is a "natural" for all poster work, theatrical as well as commercial. It is one of our extremely versatile alphabets today, as usable on a fancy letterhead as on displays, truck signs, packaging, etc. It is an easy letter to cut in film for screen process work, and one of the easiest alphabets to cut out of wall board, prestwood or wood. When cut out of heavy wood, the broad base makes it practical to stand the letter on its own base thus serving to produce a desirable silhouette or three-dimensional effect. When printed it is effective reverse or direct, and prints well on any kind of stock. You'll like to keep this alphabet handy for everyday use.

ABCDEFGHIJKL MNOPQRSTUVWXYZ

&!?$1234567890

abcdefghijklmnopqrstuvwxyz

■ ESSENTIAL CHARACTERISTICS

This novelty type face designed in 1936 is in a class by itself. Its strong identifying characteristics are: a stencil-like effect produced by disjointed stems and open counters; extreme contrast between thick and thin strokes; tall, condensed and flattened bowls. Allegro bears a remote resemblance to Corvinus in its sharp definition between thick and thin strokes, hairline serifs and compact rectangular shape. It is also somewhat of a hybrid among type faces, not only because of its relation to the Corvinus and Stencil type faces, but also because it is neither a Roman nor italic. The thick-and-thin structure is Roman in format but its slant is italic.

Another identifying characteristic of Allegro is its ball-shaped terminals which remind one of music notes. This is especially evident in the capital letters, as for example in the A, C, E, etc.

The lower case conforms to the familiar traits of the upper case. For best legibility the upper case should be used in combination with the lower case.

■ ESSENTIAL CONSTRUCTION

This is a standard type face available in a wide range of sizes, from 10 to 96 points. Where possible, especially if time is of the essence, it is advisable to use printers' type, photo lettering or any of the other ready-made lettering sheets which are now available.

For its occasional appearance on a showcard, Allegro is practical and will add a note of elegance to the appearance of the card. The extreme contrast between the thin and thick strokes makes it difficult to use one brush or one pen to accomplish this in a one-stroke technique. It would have to be rendered as a built-up letter or outlined in pen to be filled in later. If the lettering, however, is to be rendered large size, as for large exhibit panels, it becomes more practical to use Allegro more extensively. The larger the letter is, the easier it can be rendered with a single-sized brush or pen.

■ ESSENTIAL APPLICATIONS

This type face is an appropriate one to use on advertising matter dealing with the concert hall, stage or similar cultural pursuits. Allegro was not designed to be an all around alphabet intended for frequent use. Its application must be carefully matched to the occasion which calls for elegance, festivity and gaiety. It does not rate high in legibility and therefore would be inappropriate on a moving vehicle or posters which are meant to be seen at a distance. It is a fine alphabet to use when the conditions of the printing process are just right. Because of the presence of fine hairlines, the paper on which it is printed must be smooth and calendared. Printed with care on a fine glossy stock, Allegro will sparkle with the fine interplay of dark and light relationship between the strokes. Printed poorly, the fine lines will run in and disappear leaving a flocky appearance.

ABCDEFGHIJKLM
NOPQRSTUVWXYZ
&!?$1234567890

ESSENTIAL CHARACTERISTICS

The informal aspect of this type face is evident in the "incompleted" strokes joining some of the letter components. Examples are: the cross bar of the A, the uprights of the B, E, F, etc. Also, look for the overlapping element of the upper lobe of the B, the O and Q which do not quite meet, the "across the letter" characteristics of the cross bars of the A, E, F, and H. Another outstanding mark of identification is the consistent round terminals of all strokes, giving the feeling that the letters were constructed with a fully charged brush or a round nibbed pen. All letters start at an italic angle, lending vigor and a feeling of speed to this type face. Numerous hand variations may be made from this basic type face. The endings of the strokes can be square or oblique, or variations may be introduced in the thickness. Many improvisations are possible and invited!

ESSENTIAL CONSTRUCTION

A number of different lettering tools may be used to produce a Balloon type of alphabet. Each tool would introduce its own characteristics and natural modifications. If a flooded lettering brush is used, the ends or terminals of the letters will come out round. If the brush is paletted and chiseled, the terminals will be either square or oblique and the structure of the letter will take on a thick and thin appearance. A nibbed lettering pen is a "natural" as a lettering tool for this type face and will encourage speed and spontaneity. A carpenter's pencil, handled adroitly, will give this alphabet a thick and thin ribbon effect, producing a pleasing calligraphic feeling. It doesn't matter so much what modification you choose to use—what does matter is that the effect be consistent and even-toned. A careful and studious pencilling-in is not recommended even if a "reproduction" perfection is the ultimate goal. Experiment with the tool which gives you a complete stroke, rather than a thin line intended for filling in. Make several versions quickly and with full thickness—and then using your own experiments as models, select one for more careful rendering.

ESSENTIAL APPLICATIONS

The Balloon type face shown here and its close relative, Cartoon, suggest speed, a feeling of informality and a casual typographic tempo. Its original use was in "balloons" or mouth wordings of characters in cartoons and comic strips. This type face would be out of place on ads intended for institutions, products or services of a conservative nature. Balloon is appropriate for "quickie" sale signs, on occasional lines of lettering to relieve the monotony of straight type matter, as well as on headings for book jackets, direct mail brochures, bumper strip signs, etc. For best effects, use Balloon in conjunction with a contrasting type face, such as Futura, Caslon or Condensed Gothic.

ABCDEFGHIJKLM
NOPQRSTUVWXYZ
1234567890
abcdefghijklmnopqrstuvwxyz

ESSENTIAL CHARACTERISTICS

There are several formal Spencerian-like scripts cast as standard type faces. They are in addition to Bank Script, shown here, a type known as Excelsior Script, as well as Typo Script, Commercial Script, etc. Of these, Bank Script is perhaps the most widely used. All however share in some measure the characteristics which describe Bank Script, with variations that are of greater concern to the typographer than to the lettering artist for whom this model is intended.

This type, based on old formal Spencerian handwriting, is a connecting script, since the links in the lower case letters are made to join. The capital letters have a more pronounced swirl-flourish than the lower case letters. In the caps, which are intended only as initials or starting letters, the flourishes are full-bodied and possess a free-flowing grace. The strokes vary in thickness from a fine hairline for the flourishes, to a beautifully accented main down stroke, with the accented weight hovering in the center of the stroke. Even some of the thin secondary strokes, such as the rising loop in the A are gracefully accented to show modulation in pressure. Also of note are the ball terminals which appear in many of the letters.

ESSENTIAL CONSTRUCTION

There are several ways of rendering this difficult type face by hand. Each method has its own adherents and since excellence can be achieved both ways, there is no one way that is so right as to make the alternate way wrong.

There are letterers who prefer to use a fine-pointed water color brush and have developed a remarkable proficiency using the brush as the working tool. Others prefer using a flexible pen and can prove how effective their chosen tool is by the excellence of the results.

Whichever tool is used however, it must be preceded by a careful job in pencil layout. This starts not only with the conventional horizontal guide lines to establish the height of the letters, but mechanically drawn-in diagonals to assure uniformity of angle throughout.

The degree of the chosen angle is not as important as the *uniformity* of the angle. The degree to which each letter is penciled in will depend in a large measure to the degree of perfection required and obviously too, to the virtuosity and skill of the letterer. The inking in of the lettering calls for absolute control of tool, steady nerves and an ease of hand motion.

ESSENTIAL APPLICATIONS

This most formal of script letters, in application, must match the occasion for which it is intended. Bank Script (or one similar to it) is the favorite for wedding invitations, formal dances, engraved stationery or social stationery, etc. As a commercial type face, Bank Script is used but rarely, and then only for a headline for advertisement that conveys a feeling of elegance or grandeur.

Bank Script capitals cannot be used to form complete words for the same reason that capital letters in natural handwriting would be illegible. The caps add importance and start the word off with a flourish. The lower case letters carry the burden of the copy. The letters of the lower case are intended to be linked together to form unbroken chains. You cannot letter space in script, the same way as you can Roman or italic letters.

Bank Script

23

ABCDEFGHIJ
KLMNOPQR
STUVWXYZ
&.!?$1234567890

abcdefghijklmnopqrstuvwxyz

■ ESSENTIAL CHARACTERISTICS

The Roman heritage of Baskerville is evident in the graceful thick-and-thin relationship between strokes, the slender serifs and full swellings of the round elements. Baskerville closely resembles Caslon in these respects. The serifs are slightly tapered and well-integrated into the stems from which they emerge. Outstanding characteristics which unmistakably iden⸱⸱y this type face are: the long horizontal base stroke of the E which extends well beyond the top stroke; the rather low slung cross stroke of the G; the virile swash line of the Q; and the construction of the W which is formed by two V's in tandem. In the lower case, the curlecue tail of the g is an easy give-away which identifies Baskerville.

■ ESSENTIAL CONSTRUCTION

This is a reproduction type of letter and cannot be "banged out" for hurried showcards. To do justice to this type face, ruling pen and straightedge and patience are essential. In view of the fact that Baskerville is so popular as an advertising type face, layout artists need not render this alphabet by hand, but must learn to deftly indicate it on roughs and comps submitted to customer and printer. Type indication of Baskerville is accomplished with flat layout pencil with chisel just wide enough to match the thickness of the heavy stroke. The thin elements can be rendered either by using another pencil or else using the side of the chisel of the broad pencil. For a modified rendition of Baskerville on better class showcards, a flat-nibbed lettering pen can be used, with either India ink or Letterine where a gloss is desired.

■ ESSENTIAL APPLICATIONS

Baskerville is one of the most usable type faces in the typographer's repertoire. It is both a display face and a body face and as such has a prodigious range of application in newspaper and magazine advertising, book jackets, labels, letterheads, title pages and chapter headings of books and brochures and often body text as well. It is a type face devoid of frills and whims and therefore represents a very legible alphabet pretty much the same as Caslon and other basic alphabets. The uses of Baskerville on window displays and posters must be tempered by a consideration of the nature of the advertising message. For instance, Baskerville would be very much at home on a window card for an elegant haberdashery or better class department store, but it would neither be appreciated nor desirable in a butcher's window or the local hardware store.

ABCDEFGHIJ
KLMNOPQRS
TUVWXYZ
abcdefghijklmnop
qrstuvwxyz

■ ESSENTIAL CHARACTERISTICS

It is not difficult to see how the Brushscript alphabet shown here derived its name. The brush-like quality is evident in every letter of the alphabet. The general effect is one of the markings of a heavy brush used with considerable pressure on the downstroke and very light pressure on the thin elements and the connecting frills. In general, this grotesque looking type face is based more on the clumsy heavy-handed handwriting of a brute, than on the delicate sensitivity of the lettering artist. Be that as it may, it has enjoyed a wide popularity for many years. Most people I know think this to be a rather poorly designed alphabet and yet it is widely used and there is no other quite like it!

■ ESSENTIAL CONSTRUCTION

The broad strokes of the elements of Bernhard Brushscript suggest the use of a broad-nibbed instrument such as a short-haired lettering brush well palleted, a wide pen or a carpenter's layout pencil. Whatever the tool, it is used in pretty much the same way. To accent the heavy stroke, it is necessary to bear down on the tool with determination. It is the accented heavy strokes which give this style whatever character best describes it.

■ ESSENTIAL APPLICATIONS

The upper case is used only as a beginning letter of a word or as initials—never for complete words. It would not be legible if only caps were used to compose total words.

This style is seen very often in newspaper and magazine ads, book jackets, letterheads and chapter headings of books. At best, Brushscript is a type face which does not rate high in legibility; therefore it is not appropriate for poster and display work. For all the remarks I have made about this type face, it may become one of your favorites. Who knows?

ABCDEFGHIJKLM
NOPQRSTUVWXYZ
abcdefghijklm
nopqrstuvwxyz
&!?$1234567890

■ ESSENTIAL CHARACTERISTICS

Modulated grace and beauty are combined in this sophisticated alphabet designed by Lucien Bernhard.

Outstanding features of recognition are: graceful contrast between thick and thin strokes; wedge-shaped serifs; thin strokes which balance and complement the heavy strokes; hairlines and rather long serifs. Note especially the apex-pointed A, the low slung weight of the accented curvature of the G, the below-the-line simple ending of the J, the diagonal side strokes of the M and the short little V between these strokes, the overhanging diagonal of the N. Unusual too, are the connecting sets of serifs of the W and the criss cross of the two V's which form that letter.

In brief, Bernhard Modern Bold is a Roman type of letter, highly individualized and notably calligraphic in appearance.

■ ESSENTIAL CONSTRUCTION

To best remember this type face, think of it as a Roman alphabet with accented tapered hairlines and long serifs. Although this is a type face stocked by modern typographic houses and also obtainable in photo lettering styles, an acceptably good facsimile is not too difficult for the experienced hand letterer adept at brush and pen. The rather marked variation in thickness between the strokes and the shapely accents suggest the use of a flat, broad-nibbed pen, especially for casual showcard work not intended for reproduction. For reproduction purposes, color comps or actual color-on finished work, practically all strokes must be accomplished by hand, relying little on ruling pen. Close inspection will reveal that many of the strokes which at first appear straight and parallel show a faint indication of tapering; even in the upright strokes the thickness varies and thins somewhat in the center. This also applies obviously to all cross strokes and serifs.

■ ESSENTIAL APPLICATIONS

This is a "Tiffany" letter reserved for merchandise and services which cater to the elite and sophisticated. Bernhard Modern Bold would be a suitable choice for books on art and the classics, both for the cover and the jacket as well as chapter headings, title page, etc. For letterheads and labels of distinction, Bernhard Modern Bold would add to the effect and establish the elegance of any firm.

Because of its rather thin serifs and finely accented cross strokes and hairlines, this alphabet would not be a wise choice on posters, vehicular advertising or for run-of-the-mill purposes.

It is an excellent choice for motion picture and television titles and credit listings intended to appeal to a leisurely sophisticated audience.

ABCDEFGHIJKLMN
OPQRSTUVWXYZ
&!?$1234567890
abcdefghijklmnopqrstuvwxyz

■ ESSENTIAL CHARACTERISTICS

Beauty, grace, refinement. These adjectives characterize this type face designed by Lucien Bernhard. This calligraphic italic Roman is easy to recognize. Look for the gracefully wedged horizontals of the E, F, L, etc., the swash-tailed K, the high center element of the M, the virile swash of the Q and the cross over of the two V's which form the W. The lower case is unusual in the hook-like serifs of the ascenders and the rather short descenders compared with the very long ascenders. The calligraphic pen-like quality of the lower case is best exemplified in letters like the b, c, d, o, p, s and z. Note the absence of serifs in the i and l.

■ ESSENTIAL CONSTRUCTION

A flat-nibbed lettering tool is definitely suggested in both the design and the construction of Bernhard Modern Bold Italic. This is a standard type face and in stock at most of the better advertising typographers and composing rooms. The tools required will depend on the final use of this type face. The layout artist preparing a visual or "comp" will use a flat chisel layout pencil to stump in this style with sufficient character to give a preview of the final effect. The reproduction letterer working for black and white reproduction will draw out his lettering-in pencil and carefully employ a small brush for·building up. The showcard letterer may use either a flat-nibbed pen or a short-haired showcard brush to simulate this style sufficiently to capture its charm in a casual manner.

■ ESSENTIAL APPLICATIONS

An excellent type face, this, for book jackets, ads for luxury items, labels and letterheads. While a version rendered quickly in pen and ink may be used for showcards for optical stores, haberdasheries, etc., this style would not be appropriate on posters seen from a distance or on moving vehicles. This is an elegant "close up" alphabet to be seen at close range, but it is not a work horse made to carry a heavy load at full speed. The sensitive and delicate nature of Bernhard Modern Bold Italic suggests a devotion to the "finer things in life" and to markets which cater to the elegant.

Bernhard Modern Bold Italic — 31

ABCDEFGHIJKLMNO
PQRSTUVWXYZ
abcdefghijklmnopqrstuvwxyz
&.,?!$1234567890

ESSENTIAL CHARACTERISTICS

Here is an alphabet designed by one of our greatest contemporary designers who has created more than 30 type faces. His name? Lucien Bernhard. This type face which bears that famous name is called Tango and it is easy to see why. There are elegance, grace and rhythm in the controlled flourishes and the continental charm of this unique alphabet. Observe the unusual constructed A, shaped like a musical instrument. This same effect of music and the dance are shown in the construction of the B, E, F, G; in fact, in practically all the capital letters. The lower case, you will note, is more restrained and disciplined. And it is a good thing that it is, because the lower case is made to carry the burden of the message. The caps are ornamental and would be difficult to read in continuity.

ESSENTIAL CONSTRUCTION

The calligraphic quality of Bernhard Tango calls for the light touch of the pen. Look at the capital R and the Z and you will agree the pen seems to be the natural tool. This is equally true of all caps as well as the lower case. You may be adventurous enough, however, to try your well disciplined brush and make it do the work of the pen, and if that's the case, you can dance your Tango with either tool—as long as you keep up with the rhythm! For reproduction purposes, Tango is available as a photo lettering type and all you have to do is use good sense when to employ it on your layout and be able to simulate it appropriately.

ESSENTIAL APPLICATIONS

In the first place, Tango, like Old English and other decorative styles should never be used in capitals alone for continued reading matter. The caps are only for initials or the beginning of a sentence. As to the uses of this type face—there are many, but all must savor the lightness and elegance typified in the inherent construction. Chapter headings, book jackets, ads for silverware, furs and other luxury items, travel folders, etc., are some examples of where this alphabet may be used.

ABCDEFGHIJKLMNO
PQRSTUVWXYZ&
1234567890$
abcdefghijklmnop
qrstuvwxyz

■ ESSENTIAL CHARACTERISTICS

Beton Extra Bold is of the Roman thick-and-thin family of type faces and is identifiable by its square looking blocky serifs and the contrasting weights of its thick and thin elements. In general, this face has a squatty compact appearance and most letters fit into a square as shown by the H, L, etc. Letters like the M and W are considerably wider to allow space for the multiple strokes. The O and Q are ovular, rather than round, thus conforming to the rather squarish appearance of the other letters. Without changing its basic construction, Beton Extra Bold can be made into an italic. The letters may be embellished with an inline, or may take on a three-dimensional effect with the introduction of a drop shadow. See a modification of this type face, on page 37 which shows Beton Open.

■ ESSENTIAL CONSTRUCTION

Beton Extra Bold lends itself equally well to either freehand or reproduction techniques. For a freehand lettered version of Beton, use a medium length hair showcard brush, chisel-edge style. Palette each stroke carefully and do not carry too much paint. The thin elements may be done in one stroke, the thick elements will require two or more strokes. If this style is to be prepared for photographic reproduction, it is best to carefully lay out and pencil in each letter using a T-square and transparent triangle to square up all vertical and horizontal strokes. When the entire letter is pencilled-in, it is ready to be inked-in with the aid of ruling pen. First all straight lines are inked-in, after which a fine brush is used for the curved elements. When all work is thus outlined, the areas within the inked outlines are carefully filled in with a showcard or water color brush. Any irregularities are then touched up with white opaque tempera paint.

■ ESSENTIAL APPLICATIONS

Beton Extra Bold is essentially a poster letter because of its excellent legibility and boldness. It has always been a favorite type face for travel and theatrical poster work. This alphabet is also widely applicable for a great variety of other advertising purposes such as letterheads, label designs, window displays and point of purchase displays. Because Beton is structurally a simple letter form, it is popular for screen process work because it can be easily reproduced with hand-cut film stencils. For similar reasons Beton is frequently selected for three-dimensional displays and for signs. The letters are easily cut with "cut-awl." Because of its strong visual impact, Beton Extra Bold shown here is frequently used on truck lettering, railroad cars and other moving vehicles. When used in conjunction with a contrastingly delicate script, each serves as an excellent foil for the other.

ABCDEFGHIJKLMN
OPQRSTUVW XYZ
&.!?$1234567890

ESSENTIAL CHARACTERISTICS

Beton Open is a standard type face which offers a good addition to the letterers' and layout men's repertoire. It may be considered a modified Roman with square serifs enclosed within a framework of outline and shadow. The shadow case is almost the width of the basic stroke and is an integral part of the letter form. Basically the letter forms are wide to allow space for the thick shadow. The extended nature of the letter is exemplified in the B, D, K, M, etc. Note especially the X, with special emphasis on the left diagonal which forms into two sections in crossing. This construction liberty is taken to allow space for the lower half of the letter.

ESSENTIAL CONSTRUCTION

This is one case where type or photo lettering can be a blessing. To attempt to reproduce this style by hand, when the professional services of the typographer can be had for a nominal fee would surely be perverted economy. However, for a facsimile of Beton Open such as would be used on display panels, trucks, store windows, etc., the alphabet shown here will serve as a model. Though the shadow can be made in different colors, the basic letter itself has been found to look especially good if it is painted in a color lighter than the surrounding outline and shadow. For hand painting, construct the basic letter first and then add a shadow. To complete each letter with a shadow before proceeding with the next would be quite involved and often results in a lack of uniformity. Advice: Add the outline and shadow as the last step in the job.

ESSENTIAL APPLICATIONS

Beton Open is a type face shared by sign painters and typographers alike. Sign painters will find it a good letter choice for elegant truck panels, gold leaf work for banks and office doors, and for varied exhibition panels. Beton Open is a favorite among layout artists for work which is to convey a feeling of strength and elegance. This type face serves to best advantage when used in conjunction with a widely contrasting script, one acting as a foil for the other. As such it is widely seen on letterheads, labels and title pages, as well as on chapter headings in books, newspapers and magazine advertisements, direct mail pieces, etc. The effect of Beton Open can be spoiled if it appears on more than one or two lines of type. Moderation, therefore, is of the essence.

ABCDEFGHIJKLMN
OPQRSTUVWXYZ
&!?$1234567890
abcdefghijklmnopqrstuvwxyz

ESSENTIAL CHARACTERISTICS

This 18th century style gets its name from its Italian designer, Giambattista Bodoni and is considered one of the basic type faces of all time. It is characterized mostly by its fine relationship between thin and thick strokes, its engraved-like hair serifs and perfect structural symmetry. The serifs run across the stem endings without transitional fillets of any kind. The cross strokes retain the same quality and line thickness of the serifs. The axis of accent is horizontal, as you can see by looking at the O and observing the heavy accented weights of both sides deepen at their heaviest in the mechanical center of the letter form. Of special note of value in further identifying Bodoni, are the low slung cross bar of the G, the ball terminal of the J, the pin-point meeting of the two diagonal strokes of the K, the cat's tail swash of the Q and the cross over formation of the two V's which make up the W, united by a common center serif.

ESSENTIAL CONSTRUCTION

The Bodoni type face comes in several variations, such as Bodoni Bold, Italic, Condensed, etc. and in a very wide range of point sizes. The alphabet shown here is the basic Bodoni and is one of the most popular in the series. It is difficult to conceive it as any other but a reproduction letter requiring the utmost care, precision and feeling for good type design. As a reproduction letter style, the use of ruling pen and fine brush are inevitable. It cannot be "knocked out" as a quick showcard style without damage to its inherent beauty and grace.

Because this form of Bodoni, as well as most of the others in the series, is so basic, it is readily available not only as printer's type, but also in all the other mediums—photo lettering, transfer sheets, hot press type, and in pre-cut letters of wood, board, plastics and metals.

ESSENTIAL APPLICATIONS

The Bodoni version shown here is both a display type as well as a body type. This is not true of most alphabets. They are for the most part either a display face or a body copy face—and seldom can be used alternately.

As a display face, Bodoni is extremely versatile in its applications. It is a favorite for headline copy for newspaper and magazine ads, is also widely used for captions and title pages of books, letterheads, as well as packaging. Many of the higher priced department stores frequently use Bodoni for window cards and interior display boards. For window cards, individually or in limited quantities, most department store display shops have a sign writing machine which uses either metal or wood type and operates on the principle of a proof press. Bodoni is included as one of the basic fonts in these units.

ABCDEFGHIJK
LMNOPQRSTUV
WXYZ
abcdefghijklmnopqrstuvwxyz
&$?!1234567890

ESSENTIAL CHARACTERISTICS

You don't have to be the ace sleuth of the ABC detective agency to immediately spot and recognize this type face. It is a face you can't mistake anywhere. Bodoni Ultra is a display type marked by sharp contrasts. Super fine hairlines are tastefully combined with massive strokes to make this one of the most distinctive alphabets in the typographic artist's collection. Serifs and cross bars are exceedingly fine in weight and join abruptly into the main stems without transitory subtleties. Note the triangular wedged shape of the left diagonal of the A. These "wedgies" are also seen in the C, E, F, G and other letters. The outer rim of the round letters such as the C, G, O, and Q are mechanically perfect circles. The counters (inside areas) however do not follow the curvature of the outer shapes. All counters are rigidly straight and express the utmost in mechanical precision.

ESSENTIAL CONSTRUCTION

For a careful rendering suitable for reproduction, Bodoni Ultra must be made with mechanical instruments such as the compass and ruling pen. This really is not a difficult letter to make, except it takes time and care to execute it properly. Essentially there are two thicknesses. The hairline serifs, cross bars, some diagonals and uprights are made in one stroke of the ruling pen. These strokes are about 1/8 to 1/10 the weight of the heavy strokes. With the help of a marked-off piece of paper, it is a simple matter to get all heavy elements exactly the same thickness. The O and other round elements are compass-made and so actually require less control and skill than are needed to render the curvatures of the conventional Roman. After all letter forms are carefully measured off, pencilled-in and outlined with ink, the areas are filled in with ink or paint for completion of the lettering.

A modified version of Bodoni Ultra can be rendered free hand in brush or lettering pen with sacrifice of precision in favor of speed and the "personal touch."

ESSENTIAL APPLICATIONS

This is a display face and as such is appropriate for headings and titles of stories, books and advertising brochures. Bodoni Ultra can (when rendered rather carefully) be hand lettered with brush and lettering pen for use on display and counter cards for better class department stores, jewelry shops, cosmetic counters, etc. It would not be appropriate (nor perhaps appreciated) for those businesses which cater to our more basic needs. This is an alphabet which must not only be lettered slowly but it must also be viewed slowly.

For exhibit display work this alphabet is appropriate when it is lettered directly on display panels or other backgrounds. It would be hazardous, construction-wise, to attempt to cut this alphabet out of beaverboard or wood.

Bodoni, Ultra

41

ABCDEFGHIJKLM
NOPQRSTUVWXYZ
&?$1234567890

ESSENTIAL CHARACTERISTICS

This is unmistakably a thick-and-thin alphabet. The great contrast between thick and thin strokes makes this an alphabet which personifies the Beauty and the Beast. This contrast is unmarked by subtleties. There are no intermediate thicknesses. It's either extremely thick or extremely fine. Even the lobes and round parts of the letters show an unvaried thin mechanical line, as if produced with compasses. Liberties are taken in construction of cross bars which are either very high as in the E, F and K or very low as the A, P and R illustrate.

Note the side strokes of the M and the odd construction of the Y which looks like a small v with an elongated descender.

ESSENTIAL CONSTRUCTION

All thick strokes are of the same thickness and are extremely heavy by way of comparison to the thin strokes. Although this is a standard type face and has been in existence for over 25 years, its popularity as a printer's type face has somewhat faded over the years, but it is still a good alphabet for showcards. As a showcard alphabet the sample shown here represents a basic format and invites many variations in structure. For instance the thick strokes can be made thicker by way of contrast and the thin strokes can be enhanced with a slight variation in thickness to introduce some modulation in weight. As to tools—a flat well-chiseled showcard brush, a Speedball pen or a combination of both can be used. In spite of its meticulous appearance, this alphabet lends itself to speed and snap.

ESSENTIAL APPLICATIONS

This is an excellent theatrical showcard style. In addition, it fits in well with the tone of advertising associated with ladies' wear, perfumes and other feminine accessories. There is something continental about this alphabet which is in tune with art exhibits, bonbons, silverware and the like. For reproduction purposes, Broadway should not be reduced too small, especially when it is to be printed in reverse, since the fine lines may have a tendency to close in.

BROADWAY

ABCDEFGHIJKLM
NOPQRSTUVWXYZ
&.!?$1234567890
abcdefghijklmnopqrstuvwxyz

ESSENTIAL CHARACTERISTICS

This heavy-handed script belongs to the group of contemporary italic type faces which feature connecting letter links in the lower case. The lower case was designed so that most diagonal thin strokes connect up with the preceding and following letters in a series comprising a word. Another such type face is Kaufmann, shown on page 135. Brush Script bears a resemblance to Bernhard Brushscript which appears on page 27. Compare the two for structural differences.

Both related type faces show a massively weighted bold horizontal stroke produced by the well chiseled showcard brush. The down strokes produced by the edge of the chisel are thin and vary in thickness directly in relation to the angle in which the brush is held. The letters H, T and W are typical examples.

ESSENTIAL CONSTRUCTION

This is a one-stroke letter and has good potentials for speed in rendering. When produced with a felt-nibbed "magic marker" type of tool, many letters can be rendered without lifting the tool from the paper for the completion of an entire letter. With the exception of the C, G, J, Y and Z, all caps are kept within the guide lines.

While the model shown here represents the standard type as it was designed in 1942 by Robert E. Smith for the American Typefounders, extended latitude can be exercised in a showcard writer's version of this alphabet, without loss of effect. Your natural handwriting will instinctively influence your version, so that individuality will be introduced without conscious planning. Copy this alphabet as you see it here, several times, then close the book and do your own version.

ESSENTIAL APPLICATIONS

Reserve Brush Script for an occasional headline which is to convey an informal yet forceful message. This objective could easily be defeated if too much of the alphabet is used on one poster, showcard or other advertising media. It shouts so much that unless it is checked and kept to one or two words, it creates a loud confusion where nothing is heard. It is only contrast that makes for dominance. Never use the upper case only in the formation of words. Start with a cap and unite it with lower case letters as you would do in normal handwriting.

This bouncy, jazzy letter caters to the market place and appears out of place in a bank, library or temple. Keep Brush Script in its place and it will serve the purpose it was intended for by the designer.

ABCDEFGHIJKLM
NOPQRSTUVWXYZ
1234567890$
abcdefghijklm
nopqrstuvwxyz

■ ESSENTIAL CHARACTERISTICS

This beautiful type face carries all the gracious dignity of the finest in Roman type design. Each letter is individually endowed with good form, perfect balance and quiet restraint. There is a fine modulation between thick and thin strokes and the serifs blend in with the main elements unobtrusively. Of particular note are the following identifications: the lower cross stroke of the E extends well beyond the upper two cross strokes; the G is unusual because of its low cross bar; the J is much larger than the rest of the letters; and the Q is ornamental because of its graceful yet virile swash.

■ ESSENTIAL CONSTRUCTION

This alphabet must be carefully constructed to retain its identity. You cannot "knock it out" while the customer is breathing down your back. For perfect rendering in newspaper or other published advertising purposes, a ruling pen and ink technique will yield the best results. However, for gold leaf and glass work a carefully pounced pattern and deft use of the mahlstick will produce a commercially acceptable facsimile of Bulmer.

■ ESSENTIAL APPLICATIONS

Because of its inherent classic beauty, Bulmer must be reserved for things connected with gracious living and the arts and the softer sentiments. It is appropriate for book jackets dealing with romantic novels and the classics. Reproduced by silk screen or other mechanical means, this alphabet is suitable for department store windows, banking institutions and the like. Because of its comparatively thin hairlines, it should be reproduced with the proper mechanical make-ready to avoid blurs in the printing operation.

ABCDEFGHIJKLMN
OPQRSTUVWXYZ
1234567890
&
!?

ESSENTIAL CHARACTERISTICS

A type face of rare calligraphic beauty, Carolus represents an ideal marriage between lettering and calligraphic writing. It shares with all calligraphic art the following identifying features. The round elements are ribbon-shaped and show the decided twist of connecting strokes. No attempt is made at artificial refinements in modulating weight gradually, from thin to thick. The D is a typical example of this characteristic. Then too, you will note that the letters are rather wide and somewhat square-shaped, as exemplified in the letters O and Q. Careful inspection will reveal the presence of fine spur serifs which are not symmetrical; the top serif faces to the left, the bottom counter part faces right. Note too that a lower case U is used. Also, if you look closely you will see a slight bend in the diagonal strokes, as in the letters A, M, N, V, etc.

ESSENTIAL CONSTRUCTION

In the opinion of a true calligraphic artist, it would be an act of sacrilege to render this style with anything but a proper calligraphic tool. This may be a broad-nibbed well chiseled lettering brush or a chisel-edged pencil cut at an angle. The twists and abrupt turns, the spur serifs, the slight bend of the vertical stroke—all would be the direct result of the character of the tool and would be the only honest presentation of this true-blue calligraphic letter form. Don't, *don't* try to artificially arrive at this true calligraphic nature of Carolus by laboriously outlining each letter with thin lines and superficially copy that which comes so naturally when the right tools and techniques are combined with a real understanding and appreciation of this beautiful letter form.

ESSENTIAL APPLICATIONS

Carolus, for all its natural beauty of line and form, has limited uses in the commercial repertoire of the sign and showcard writer. Like most calligraphic type faces, Carolus lacks sufficient boldness to make it acceptable for the general run of posters and displays. To be properly appreciated, this lettering style must be viewed at close range; this makes it appropriate for book titles or covers and title pages—especially literature dealing with subjects of the arts.

The calligraphic flavor which characterizes this type face makes it also a good alternate for the more ornate forms of Old English alphabets which are traditions for diplomas and certificates. Carolus is more legible than Old English, and can generally be rendered more quickly.

Try your hand in duplicating the spirit of this letter form, but don't try to copy the example shown here with a counterfeiter's care.

CAROLUS

ABCDEFGHIJKL
MNOPQRSTUV
WXYZ &!?.:!?$1234567890
abcdefghijklmnopqrstuvwxyz

■ ESSENTIAL CHARACTERISTICS

Caslon is one of the most beautiful of the traditional alphabets in the typographer's gallery of type designs. The classic beauty of Caslon has made it an all time favorite and one of the best exponents of the family of Roman alphabets. Caslon is distinguished by its cut-in-stone monument quality which distinctly shows traces of the mark of the chisel in the crispness and contrast of its thick and thin elements. The thin parts give evidence of the work of the stone cutter's tool used sideways while the broad strokes show the effect of the full face and accented strokes of the tool. Other recognizable features are: the extension of the A with its oblique cut off top; the overlapping diagonal stroke of the N; the full blown D, O and Q, each with a delicate balance between thick and thin; the W which appears to be composed of two intersecting V's; the beautiful flourish of the tail of the Q.

■ ESSENTIAL CONSTRUCTION

It takes a lot of time and talent to do creditable hand lettered reproductions of the standard Caslon type face. In view of the fact that this popular style is readily available in a great variety of type sizes as well as in paste-up and photo lettering, it becomes questionable whether an attempt should be made to meticulously recreate by hand that which is so abundantly available in ready-made forms. However, where some special purpose calls for a hand lettered Caslon, the letter forms should be carefully pencilled-in and meticulously worked over before the inking stage is reached. All straight lines can be ruled-in with ruling pen, while the outer rim of the C, G, O and Q can be compassed-in. All other work can be brushed in either with a standard chiseled edge, round ferruled showcard brush, or if preferred with a pointed artist's water color brush. When the style is used for non-reproduction purposes, such as for showcards, posters, etc, Caslon can be simulated either with a square-nibbed pen or with a well paletted showcard brush, single stroke or built-up.

■ ESSENTIAL APPLICATIONS

Caslon is available in many type sizes, both in capitals and lower case, as well as in italics. In its larger sizes, the caps can be used as a display face, though the lower case is often used for display as well. The most frequent use of Caslon is for straight type matter such as in books, body texts in newspapers and magazine ads, direct mail, etc. Though Caslon is occasionally seen on posters, it does not have sufficient carrying power when seen from a distance, though it is one of the most legible type faces for all work seen at close range. The absence of frills makes Caslon a good choice for all purposes which are designed to convey a feeling of stability and tradition. It is for this reason that Caslon is a good letter to use on signs and window lettering in color or gold and silver leaf for banks, institutions and high class stores.

Caslon No. 540

ABCDEFGHIJKLM
NOPQRSTUVWXYZ
abcdefghijklmnopqrstuvwxyz
&!?$.:1234567890

ESSENTIAL CHARACTERISTICS

This type face is one of our country's favorite type faces. It was developed by L. B. Benton who was one of the top designers of the American Typefounders for many years. It has grace, a classic beauty, is rather high in legibility and its versatility is such that it is used both for headline and display as well as for massive body copy.

A true Roman serif letter, Century has balance and symmetry. The relationship between the thick and thin strokes is consistent throughout the alphabet. The serifs are flat at the base line and blend softly into the stems. The weight of accent of the round letters is in a horizontal axis, the heaviest weight being in the center of the curve.

Identifying letters are the short center cross bars of the E and F, the below-center cross strokes of the G, the ball terminal of the J, the virile swash stroke that crosses the interior of the Q and the weighted swash tail of the R.

In the lower case the formation of the g is an identifying element.

ESSENTIAL CONSTRUCTION

Century Expanded is available in the widest possible assortment of point sizes as well as in photo lettering systems and ready made cut-out letters. The occasion often arises where the letters have to be drawn by hand for layout, pencil outline in cut-awl work, or for direct brush-in onto a painted panel or other surface. It is of practical importance therefore to duplicate a facsimile of this type face by hand. Points to remember: The serifs which touch the upper and lower guide lines are flat and run along the guide lines. They taper however, forming a fillet where the serif meets the main stroke. All of the heavy strokes are of uniform thickness with the exception of the widest part of the accented weight of the curves, which are made a shade heavier.

For a more accurate rendition of this type face, the pencil layout should be carefully executed. If the letters are to be rendered in large size, it would do well to actually measure off the thickness to assure uniformity. The time spent in a carefully pencilled job will pay off in the quality of the finished work.

ESSENTIAL APPLICATIONS

The versatility of this type face is attested to by the fact that the caps are so often used for display purposes, while the lower case (combined with caps) is one of the "standby's" of many books published for school and general use. Century Expanded as a body type has enough stamina to print well on most paper stock and is extremely legible and restful on the eyes.

The lower case when used in advertisements, is employed sparingly as secondary copy matter. The caps are extremely functional for magazine and newspaper ads, T.V. titles, labels, stationery, packaging, etc. Though Century has a distinct personality as a type face, it is proper to use it for advertising an unlimited range of products and services, blending well in any surrounding.

ABCDEFGHIJKLM
NOPQRSTUVWXYZ
abcdefghijklmnopqrstuvwxyz
&.!?$1234567890

■ ESSENTIAL CHARACTERISTICS

In most cases where a Roman typeface has an italic, there is very little difference between these two forms other than one is a straight up and down, while the other is on a slant. While this is also true of the Century Expanded caps, the lower case varies so widely between the Roman and Italic as to virtually constitute an entirely different type design. Referring to the Roman version of this alphabet as it appears on page 53 you will see the sharp difference between the Roman and the one shown here.

As a point of comparison, observe the f. The stem of the Italic f is finial shaped and weighted in the center of a stroke that rises below as well as above the body line. Note the graceful low line start and finish of many letters, typified in the i. Also study the finely tapered center of the main strokes of the i, m, n, u, etc.

■ ESSENTIAL CONSTRUCTION

Although the construction of the caps does not differ appreciably between the Century Expanded Italic and its parent Century Expanded Roman, the lower case of the italic variation has been basically altered. We will therefore focus our attention on the lower case exclusively.

The character of lower case Century Expanded Italic is determined greatly by the way the short stems of letters like the i, m, n, etc. are constructed. You will note that these stems are not of uniform thickness throughout the stroke. The letter i will serve as a good example. A fine and full swing which starts the letter leads into a tapered down stroke which as it approaches the bottom makes a subtle turn in direction and emerges as an upward swing, to balance the upper bend in stroke. Actually the two edges of the short body stroke are not parallel to each other except perhaps momentarily in the center of the stroke.

Using this analysis as a basis, you will have a guiding principle for rendering many elemental strokes of the lower case.

■ ESSENTIAL APPLICATIONS

The lower case of Century Expanded Italic is one of the most popular italics in use today. While it has been widely adopted by printers of books as a basic body type where italics are needed, it is also extremely serviceable for the typographic layout artist for a broad range of advertising purposes. You'll see Century Expanded Italic caps used for headlines in combination with its lower case. The lower case in smaller point size is widely used as secondary as well as body copy.

The example shown here will serve as a model for lettering and display artists who want to introduce a typographic effect which departs from the usual letter styles which have become trade clichés in the field of sign painting and showcard lettering. Century is an "advertising agency" letter style, ideally suited for display and point of purchase advertising projects for better department stores, industrial exhibits, etc.

ABCDEFGHIJKLMN
OPQRSTUVWXYZ

abcdefghijklmnopqrstuvwxyz

&!?$1234567890¢;:

■ ESSENTIAL CHARACTERISTICS

The name of this type face is derived from its obvious similarity to the letters chiseled on stone monuments. Basically it is a Roman letter form with an open face inline. This creates the illusion of the letters being raised or depressed, depending on how you interpret it at the moment. The serifs are wedge-shaped and uniform in all letters. These serifs add to the illusion of a letter produced by hammer and chisel, guided by the deft hand of a calligraphic sculptor. Although Chisel was cut as a type face in 1939, it has remained in a distinct class by itself and cannot be mistaken for any other type face.

■ ESSENTIAL CONSTRUCTION

Let's analyze the letter E to serve as a guide for construction of the others. The structure consists of a thin outline within the format of a full-bodied Roman letter. Notice that the right side of the outline is heavier than the left. Also note that the thin skeleton outline letter is shifted to one side of its broader casing. The Chisel alphabet is a tedious letter to render by hand. It is shown here not so much as a model to copy, but rather as a type specimen available to the typographic layout artist.

■ ESSENTIAL APPLICATIONS

Chisel is an attractive novelty display type face which can be used to express a particular mood or effect. Used with discretion this alphabet can add a note of classic poise and conservative elegance to a typographic layout. It can be used for chapter headings for books, headings for ads or any other one line copy. As a lettering style, Chisel would also be practical for gold leaf window lettering in banking institutions, libraries and institutions of learning.

A B C D E F G H I J K L M N
O P Q R S T U V W X Y Z
1 2 3 4 5 6 7 8 9 0
a b c d e f g h i j k l m n
o p q r s t u v w x y z ! ?

■ ESSENTIAL CHARACTERISTICS

This is a modern square-shaped Gothic type which reflects today's emphasis on engineering and technology. It is a standard single-thick printer's type which has a unique serif characteristic. This is one of its outstanding marks of distinction. The serifs which are slab-like in construction are one-sided in letters such as A, H, K, M, etc. This is not a uniform characteristic since in some letters the serif is bi-symmetric and runs across the stem of the letter in perfect balance. Another unique feature of City Light is the lack of consistency in the matter of elbow curves of the letter forms. This is typified by the letter C, which has rounded elbows on the left side and sharply squared-off elements on the right side. Also note the fore-shortened V-shaped center element of the M, yet the center element of the W is conventional, lining up in full height to the top of the letter.

■ ESSENTIAL CONSTRUCTION

All elements of this type face are of equal thickness. This includes all strokes of each letter as well as the thickness of the serifs. It is a comparatively simple letter to render either for showcards or reproduction styles. For showcards and other related uses, the letters can be constructed in one stroke with a well paletted short-haired rigger. If a pen is used, the nib must be flat and in width match the thickness of the stroke, so that no build-up strokes will be necessary. For reproduction purposes, this is an ideal letter for ruling pen and straightedge since the only curvatures are the elbow corners. This also is an ideal letter to render roughly for type indication purposes on layouts, since it lends itself admirably to the use of a chisel-edged pencil.

■ ESSENTIAL APPLICATIONS

The mechanical nature of this type face suggests the clean cut rigidity of products or services dealing with machines, technology and precision. This type face has been adopted as one of the "standard" faces used by advertisers such as IBM, important firms associated with steel, architecture and the electronics field. Its uses, however, need not be confined to these specialized areas, because its clean lines and easy legibility make City Light applicable to department store posters, exhibition work and in general advertising for newspapers and magazines. It is also a popular lettering style for silk screen since it reproduces well when stencils are made by hand-cut film, photographic or even paper stencils. What is more, the absence of hairlines or fine serifs contributes to its sharp printing quality for either direct or reverse printing.

ABCDEFGHIJKL
MNOPQRSTUV
WXYZ
abcdefghijklmn
opqrstuvwxyz
&!?$1234567890

■ ESSENTIAL CHARACTERISTICS

Clarendon shown here is a slab-serif Roman. This refers to the massive square-cut serifs which comprise an integral part of the letter structure. Each serif, which bears the same weight as the thin part of the letter, sits firmly and squarely within the guide line and adds a great measure of structural strength to the letter. There is a mechanical consistency about the relationship between the thick and thin elements of the letter. This relationship does not vary. Of special note in the aid of identification of Clarendon are the following: the shortened center of the E and F which seems compactly tucked in with the other horizontals; the ball-shaped kern of the J which does not fall below the bottom guide line; the three slab bars of the W which form a strong horizontal finish to the top of the letter; the upswing tail of the R and the powerful swash of the Q which cuts through the letter with virility and determination.

■ ESSENTIAL CONSTRUCTION

This is somewhat of an expanded letter form and care must be taken to provide enough space in the layout of copy. This especially is true because the serifs, too, are rather long and demand comfortable space so that the serifs of one letter do not contact those of another. Because of the versatility of this alphabet, it is frequently used as a reproduction letter and type face for printing, as well as a hand-painted letter for posters, displays, etc. It is a comparatively easy letter to construct for reproduction purposes, because it permits the use of ruling pen and straightedge. The curves or lobes, however, have to be rendered freehand, either with pen or brush. The numbers especially call for the expert's use of freehand rendering since the greater portion of the component strokes cannot be achieved with compass or ruling pen.

■ ESSENTIAL APPLICATIONS

This is one of today's most popular type faces. It has legibility, stability and impact—all important requisites for advertising and display work. If you visit industrial exhibits, you will see numerous examples of Clarendon on posters, displays and signs. You will see evidences of it on the hand-painted form as well as three-dimensional versions of wood, plastic and stainless steel. In books, Clarendon is often used for book jackets, chapter headings, title pages. And you will recognize Clarendon on TV title and credit flashes.

Keep Clarendon handy for ready reference. You'll find you'll be using it often.

Clarendon Bold

A B C D E F G H I J K L M N

O P Q R S T U V W X Y Z

!?$1234567890¿¡

abcdefghijklmnopqrstuvwxyz

ESSENTIAL CHARACTERISTICS

This alphabet is in the tradition of formal penmanship at its best. The most outstanding characteristic of Commercial Script is the gracefully modulated variation of thick and thin strokes. The heaviest part of the accented strokes is in the center. The thin strokes are hair-fine pen lines. The general effect is a pen quality with accents due to pressure and flexible nib of the pen.

ESSENTIAL CONSTRUCTION

For reproduction purposes, this type will not allow itself to be rushed. The lettering must be well laid out in fine pencil indicating all the accents and curves. A well rendered job calls for painstaking labor with fine crow quill pen, handled with extreme accuracy. The lower case permits use of a ruler for many of the ascenders and descenders. A slight modulation in thickness is evident even in the thin hairline strokes. The pen must be kept free from grit so that it will not pick up particles of dust or other foreign matter. A smooth cardboard is essential, free from lint or surface textures.

ESSENTIAL APPLICATIONS

The capitals may be used as decorative initials in themselves or with the lower case. They should not be used to form words in combination. Words formed only with capital letters would be difficult to read. Though this alphabet is an appropriate type for engraved stationery and other highly dignified typographic effects, a hand lettered version may be developed for more casual uses for window cards for department store advertisements featuring fine furs, fine furniture, silverware and other "par excellence" features.

ABCDEFGHIJKLM
NOPQRSTUVWXYZ
&!?.$1234567890
abcdefghijklmnopqrstuvwxyz

■ ESSENTIAL CHARACTERISTICS

Contact Bold Condensed shown here is a basic alphabet without frills and doodads. This alphabet is essentially a thick-and-thin style with simple blocky serifs. In its condensed form, all traditionally round shapes are flattened as is evident in the lobes of the B, C, D, etc. Also note the following identifying features: the center strokes of the E and F are shorter than the top and bottom strokes and the serifs are somewhat thicker as well; the J ends with a ball terminal and stays within the bottom guide line; the V shaped element of the M extends to the bottom; the swash stroke of the R generally expected to take a diagonal direction is here condensed and made straight and parallel to the main upright stroke.

■ ESSENTIAL CONSTRUCTION

This is a popular alphabet. So much so, that it is available in many media—printer's type, photo lettering, pre-cut three-dimensional forms of wood, board and plastic. For reproduction or exhibit purposes, it would be more advantageous to avail oneself of these facilities rather than to take time to laboriously duplicate by hand that which is obtainable cheaper and better in ready-made form. For hand-painted showcards, truck lettering, painted panels, etc., this sample of Contact Bold Condensed will serve as a guide. For casual showcard work, a good facsimile can be rendered with a well paletted short-haired showcard brush. The paint should be of rather heavy consistency so that the square-cornered serifs can be accomplished with a natural stroke of the brush. The newly developed steel "brush pen," should prove useful for rendering this alphabet in ink.

■ ESSENTIAL APPLICATIONS

As with most basic alphabets, this type face is versatile and has many uses. It would be considered an appropriate selection for general newspaper and magazine advertisement, would go well for book jacket and title page layouts, poster advertisements, letterheads, etc. Needless to stress, the total effectiveness of this type face would depend upon other factors such as color, illustrative material and harmony with other type faces in the layout.

ABCDEFGHIJKLMN
OPQRSTUVWXYZ
&!?$1234567890
abcdefghijklmnopqr
stuvwxyz

ESSENTIAL CHARACTERISTICS

No type face in contemporary history has made so strong a "comeback" as Cooper Black. Designed about 40 years ago by Oswald Cooper, it never became quite extinct, but lay dormant until about 1961 when it was revived and became more popular than ever before. At first it was seen quite infrequently and although it was somewhat of a favorite among showcard writers, layout artists and art directors. Today there are not too many display type faces which compete with Cooper Black in popularity.

The principal characteristic of Cooper is its chubby, clubby serif structure. The serifs are globular and blend into the strokes as integral elements of the letter structure. The serifs are short and rounded like the bottom structure of a rocking chair. Other unusual characteristics are the diagonal inside areas (counters) of the Q and O. The lower case retains the same serif characteristics. Note that the ascenders and descenders are very short. Also note that the heavier part of the round elements are very small and almost appear as if they have a tendency to close in.

ESSENTIAL CONSTRUCTION

The best way to achieve the well-rounded effect so typical of Cooper, is to use a flooded brush, charged with a paint of rather loose consistency. The paint should "flow" from the brush. If the brush is heavily charged with paint of the proper consistency, the serifs will naturally swell producing the rounded cushion effect which is the chief characteristic of this interesting type face. Though we speak of a heavy free flowing brush, Cooper lends itself admirably to pen work as well. A round-nibbed pen is recommended for economy of stroke and desired effect.

ESSENTIAL APPLICATIONS

Judging from the contemporary uses of Cooper, it seems to be popular with nearly every type of ad, poster or display. It is much in evidence in magazine advertising. It is interesting to note, that not only is Cooper so much in use today, but is generally shown in very large type sizes, apparently out of proportion to the page. Evidently the aim is to achieve a massive effect, overpowering in its visual impact. Cooper Black is no soft spoken type face. It fairly shrieks and pushes all other type faces off the page with an audacity in both blackness and size. Look for examples of this revived type face, once relegated mainly to the showcard writer and poster artist. You'll see it used for ads for coughs and cars, travel and toys, books and bonds. We don't know how long this popularity will last; in the meantime, get on the bandwagon—with Cooper!

ABCDEFGHIJKLM
NOPQRSTUVWXYZ
&!?$1234567890

■ ESSENTIAL CHARACTERISTICS

Although this is a standard type face, traditionally accepted as one of the basic letter styles, it is at present more in vogue than ever before. Why? There is a current trend toward extended alphabets such as Venus Extended, Latin Wide, etc. Whereas the general trend toward extended type faces may in time give way to other typographic vogues of the day, the Copperplate shown here is so basic that it will no doubt retain its popularity for a long time to come. In essence it is a one-thickness simple Gothic, stripped of all embellishments. Its chief characteristic is the extremely wide structure, finished off with the slightest suggestion of serifs. Note also that the center cross bars of the E and F are short; the O is elliptical while the Q is round.

■ ESSENTIAL CONSTRUCTION

This is such an easy to construct alphabet that one careful perusal of the specimen shown here, should imprint an image of it in the mind of any experienced letterer. It may be rendered freehand for casual use or with ruling pen for reproductive purposes. For a freehand rendition, a short-haired well paleted chisel showcard brush will naturally yield to the one thickness construction requirements called for. The wispy serifs are the results of a deft flick of the brush at the terminals. For reproduction use, the serifs may be achieved with a crow quill pen after the letters are completed with ink and ruling pen.

■ ESSENTIAL APPLICATIONS

Though Copperplate Gothic was originally designed as a letter intended for engravers' use in fine stationery and in small point size, the type face has been adapted for display work, space advertising in newspapers and magazines and some types of posters. This type reproduces well in direct or reverse printing; is extremely legible and easily read even at a distance. It is for this reason that it is used frequently on street and road markers and signs the world over.

COPPERPLATE GOTHIC BOLD

ABCDEFGHIJKLMNOP
QRSTUVWXYZ
$1234567890
abcdefghijklmnopqrstuvwxyz

■ ESSENTIAL CHARACTERISTICS

Corvinus (pronounced Cor-veenus) is an ultra sophisticated lettering style with distinct characteristics that identify it immediately. Here are some unmistakable clues. The cross bar in the A is unexpectedly thick compared to the thin diagonal or extremely fine hairline serifs. Look for the same feature in the cross ties in the E, F, H, L, T, and Z. All lobes (commonly round parts) are flat, as for example the B, D, O, etc. The hairline appendage which acts as a sort of serif on the upper part of the C appears also in the G and S. The U is constructed along the lines of a lower case, and the serif appears only at one side of each upright stroke. The tail of the Q points straight down. The M is composed of diagonals only—no straight up and down strokes. Can you find other tell-tale characteristics?

■ ESSENTIAL CONSTRUCTION

This is definitely a reproduction letter and even a fair hand painted reproduction requires exacting care. If a hand drawn version is called for, the letters must be very carefully drawn in first, before ruling pen and straightedge are put to use. The hairline serifs are precision made, and unintentional variations will show up disappointingly in the finished effect. Since this alphabet, long a favorite among typographers, is obtainable in type as well as photo lettering, it certainly would appear to be an unnecessary feat to render a black and white reproduction by hand. Where this type is to appear in color, as for example, on a large display panel (and neither photo lettering or type can be used) a hand painted facsimile of Corvinus can be rendered using the alphabet shown here as a guide.

■ ESSENTIAL APPLICATIONS

There is every indication that Corvinus will remain with us for a long time. It is not a passing fad and therefore deserves a place in your book of standard type faces. Where can you use Corvinus? Look through the newsstand magazines and you'll find it used frequently in story heads as well as in ads. Corvinus is popular for letterheads, packaging, and direct mail pieces of all sorts. It is not the best type of letter to use on posters and trucks because the thin lines seem to fade out in the distance or in motion. It is not especially good for reverse printing because heavy inking causes the thin lines to smudge and close up—especially when the type is reduced to a small point size. Where it is used with discretion, Corvinus will lend an air of distinction to any good piece of printing.

ABCDEFGHIJKLMNOPQRSTUVWXYZ

&!?$1234567890

abcdefghijklmnopqrstuvwxyz

ESSENTIAL CHARACTERISTICS

One of today's bright new type faces, Craw Modern Bold shows a clean-cut contrast between thick and thin elements. This marked contrast between hairline thinness and bold main elements gives the Craw alphabet shown here a feeling of engraving with a fine precision instrument. All letters are rather wide for their height, and each fits into an imaginary square. The serifs are uniform in thickness—or rather thinness—for they are fine and long and help the letters line up horizontally as seen in made-up words. Incidentally, the serifs are of the same weight as the hairlines and cross bars, as you will clearly see when you examine the H.

The lower case retains the same characteristics as its big brother in the sharp definition and relation of thick and thin strokes and hairline serifs. The ascenders and descenders are comparatively short thus helping to give the letters a squatty effect with a minimum of "bounce" of ascender and descender stems.

ESSENTIAL CONSTRUCTION

This is not a "bang out" alphabet which rolls off the brush just like that. In fact, Craw is essentially a reproduction lettering style which needs careful layout and meticulous rendering to retain its pristine engraved look. For reproduction purposes, a ruling pen will be needed for the outline of the straight letter strokes and a fine-haired lettering or pointed water color brush for the curved elements. Its current popularity is attested to the fact that Craw is readily available at most type houses. It also can be found in most photo lettering sample books, as well as in the specimen page of paste-up lettering sheets. In addition to the model alphabet shown here, this type face is also available in several modified versions which include a condensed style where space conservation is indicated on the layout.

ESSENTIAL APPLICATIONS

Craw is used very frequently in advertising layouts for newspaper and magazine work. Since it is a letter devoid of eccentricities or frills, its use is widespread for a very large range of product merchandising. In addition, it is a favorite style for captions and chapter headings of books, manuals, annual reports and the like. Because of the hairline serifs and cross strokes, care must be taken not to reduce the size too much, especially when intended for reverse printing. The chances are that the printer would encounter considerable difficulty in keeping the thin lines from closing in. Have you ever seen a poor job in reverse printing when a type face is used which has thin lines? The words come out with parts missing—as a sort of stencil effect which makes reading difficult. Remember that it is not how nice the layout looks on the drawing board that counts. It's how the finished work looks when it's printed that really can make or break the reputation of both printer and type selector.

ABCDEFG
HIJKLMN
OPQRSTU
VWXYZ

■ ESSENTIAL CHARACTERISTICS

Here is an interesting variation of the Roman type face. It retains the family resemblance to Roman in its thick and thin structure, with long well tapered serifs added. The outstanding characteristic of Cristal is, of course, the double stroke inline which is placed on the heavy strokes. This gives the type face a sparkling appearance and no doubt has influenced the name this type face carries.

■ ESSENTIAL CONSTRUCTION

As in all alphabets which are distinguished by an inline or other embellishment, it is best to lay out the letter as a solid first and then add the decorative element after the layout is completed. In the case of Cristal, the essential basic structure is a Roman with serifs finely tapered. Special letters to watch out for are: the open joining of the two lobes of the B; the diagonals of the K, which do not touch the main stroke; the splayed sides of the M; and the open lobes of the P and R. In normal practice, liberties may be taken with the structure of the letters and yet not sacrifice the sparkling glossy effect.

■ ESSENTIAL APPLICATIONS

Since this type face suggests a glossy effect, it is a "natural" for posters and other printed matter dealing with glassware, dishes, aluminum and other products of a metallic or glossy material. These are the more obvious uses of this type face, but the application of this alphabet need not be limited to the obvious. The creative artist, be he a letterer or illustrator, goes beyond the commonplace and finds new uses and applications for graphic art forms.

ABCDEFGHIJKLMN
OPQRSTUVWXYZ
&!?$1234567890

ESSENTIAL CHARACTERISTICS

This is a decorative inline alphabet which expresses classic dignity. Basically it is a Roman letter form with a mere suggestion of a serif. In addition to the consistent inline which divides the basic strokes equally, the most recognizable letters in this alphabet are the A with its extending apex diagonal, the extended D, the letter J with its rather narrow hook and the M with its classic diagonal strokes. Also rather unique are the open lobe of the letter P, the cat-tail extension of the Q and the lower case type U, used here as a capital.

ESSENTIAL CONSTRUCTION

The best way to render this alphabet most expediently by hand is to carefully construct each letter as if it had no inline at all. After completion of the solid type letter, the inline is carefully added with ruling pen or freehand, depending on the degree of accuracy required. The outside perimeter of the O and Q are perfect circles, while the inner shapes are ovals. This is a standard type face and where no deviation is wanted and sharp black and white copy is needed, either printer's proofs or photo lettering paste-ups are available.

ESSENTIAL APPLICATIONS

The classic cut stone effect of Delphian would lend dignity to any art work, advertising books on art, literature and music. This is the kind of lettering that would look good on the windows of a bank, or inscribed in cut stone on the frieze of a library or museum. It would be ineffectual to adopt this alphabet for routine showcard work. It is a good alphabet to include in your growing collection of type faces to be used where special effects are desired.

ABCDEFGHIJKLMNOPQRS
TUVWXYZ
&!?$1234567890
abcdefghijklmnopqrstuvwxyz

ESSENTIAL CHARACTERISTICS

Dom Casual is typical of the modern informal type designs modelled after lettering made to look as if it were brushed in by the average person doing his own sign. The casual nature of this type face is evident in practically every letter of the alphabet. There is a complete absence of strained rigidity—no signs of the marks of a ruling pen or mahlstick. The round elements (or those conventionally round) are somewhat wedge shaped, as you will note in the lobes of the B, D, P and R, as well as in many letters of the lower case. Although this is essentially a one-thickness letter, there are variations, principally around the curves, producing a ribbon-like effect which twists and turns as the direction of the stroke dictates. Note the horizontal strokes of the E, F, L, T and Z which are somewhat wedge shaped and appear to take an uphill sweep. Observe also the low slung S in both the upper and lower cases so typical of all free styles of lettering in this category.

ESSENTIAL CONSTRUCTION

The best way to do this alphabet is to analytically study the model shown here long enough to understand the effect and then to do it from memory, later to compare your rendition with the example here. It is not intended that this model be slavishly copied; rather that the effect be duplicated. Here are some hints: hold the lettering brush rather rigidly—almost as if you were handling a pencil. Do not twirl the brush between the fingers as you would normally do when doing conventional lettering. Normally when you letter you change the angle and the direction of the brush to correspond to the particular stroke. Not so here. Keep the brush at the same angle whether you are making a horizontal, angular, vertical or round stroke. Keep experimenting with the brush as you would with a flat chisel layout pencil and you will achieve an effect similar to the one here, although the letters may differ. Perhaps it is better if they do.

ESSENTIAL APPLICATIONS

Dom Casual is a casual letter, not a sloppy one. There is a very fine aesthetic line of demarcation which differentiates the one from the other. This is a very modern alphabet widely used in magazine and newspaper advertising as well as on posters, car cards and window displays. Caution: Do not use Dom Casual to excess. One or two lines of copy ordinarily suffice to draw attention to a focal headline or feature, provided the main body of the copy is carried in a more conventional type face. Let this style act as a foil in contrast to some basic alphabet. Because this is an exceptionally easy letter to cut in film, it is a popular and most welcome type for the screen process stencil cutter. It looks good direct or reverse and because there are no fine hairlines, Dom Casual "wears" exceedingly well for all types of printing processes.

ABCDEFGHIJKLM
NOPQRSTUVWXYZ
abcdefghijklmn
opqrstuvwxyz
&!?$1234567890

■ ESSENTIAL CHARACTERISTICS

This alphabet, a modern version of a 19th century type face has gained in current popularity among typographic layout artists. It has stamina, directness and good legibility.

This type face is in the family of slab serif letter forms. The serifs are not merely accidental appendages caused by a flick of the lettering tool; they are very definitely an integral part of the letter. The serifs are more than average in length and are square ended and of uniform thickness throughout most letters of the alphabet. The exceptions are the right hand side of the E and F where the serifs change into a block. This is also evident in the C, G, T and Z.

The letters are relatively wide for their height, thus creating a stocky compactness which looks well in word combinations.

The lower case bears the identical characteristics as the upper case. Note the very short and stocky ascenders and descenders, all helping to retain the compact structure of the alphabet.

■ ESSENTIAL CONSTRUCTION

This is a comparatively simple letter to render, once the relationship of thick and thin strokes are established. The serifs are wide slabs which anchor the letters to the top and bottom guide lines. In the formation of words, letters may be placed close together with the serifs of one letter almost touching those of its companions. Because of the extended form of this alphabet, care must be taken to allow a lot of horizontal space since the letters are much wider than they are high.

This is an ideal reproduction letter since most strokes are mechanically straight edged, allowing the use of ruling pen. The round elements of course must be rendered freehand because they are not segments of a circle.

Though both upper and lower case are shown here, the upper case is used most extensively for display work and point of purchase advertising.

■ ESSENTIAL APPLICATIONS

Egyptian Expanded is especially appropriate as a display letter and is used as such in newspaper and magazine advertising, as well as in poster work. Its boldness and legibility make it a popular alphabet for lettering to be seen from or on moving vehicles. Many railroad passenger cars are inscribed in this alphabet, the letters forming a long continuous frieze, the long serifs visually uniting the letters as the train speeds by. It is a good alphabet to use for truck lettering as well.

Aside from these heavy duty tasks, Egyptian Expanded is very much at home as a modern typographic motif for letterheads, labels and business forms. It gives a clean precise image to a line of type and is especially effective when the copy is kept to one long line. It mixes well with other styles.

A B C D E F G H I J K L M N O P Q R S T U V W X Y Z

& $ 1 2 3 4 5 6 7 8 9 0 . , ; : ? !

a b c d e f g h i j k l m n o p q r s t u v w x y z

ESSENTIAL CHARACTERISTICS

There are many varieties of "Old English" alphabets but only a few are standard printers' type faces. The others are for the most part hand lettered and since the opportunities for variations are so numerous (and tempting) it is good to occasionally go back to a standard face as a self-disciplining measure.

The Engravers Old English type face shown here may be used as reference. The essential difference from others in the general group is that the thick part of the strokes, as in the left hand side of the B is really composed of 2 strokes parallel to each other. This also pertains to the F, I, J, K, L, R and S, but is not consistently followed. Compare these letters to the A or C. In some instances a thin vertical hairline is used to accent the weight as in the D, E, H, etc. In addition, 2 small hairline cross bars are used to bridge the thin vertical to the right side accented element, as in the D.

ESSENTIAL CONSTRUCTION

A flexible broad-nibbed pen with a diagonal chisel-cut edge is the most likely tool to produce the desired results. However it is not the pen alone (or any other tool) that guarantees success. It is the hand that guides its direction and pressure that is most important. Difficulties will be encountered at first in keeping the fine lines as thin as they should be, the temptation arising in the mind to use 2 different pens, one for the heavy lines, one for the hairlines. That would not be the professional way and it should be avoided from the start.

The caps take longer to do than the lower case because of the multiplicity of straight lines and curves. The lower case letters admit for rapid rendering. Each stroke counts, no stroke is repeated. With experience not only will speed increase, but a rhythm will develop which will add to the character of the finished product.

For word formation, caps may be used only as initial letters, never for complete words. It's the lower case letters which are the real working members of this typographic family.

ESSENTIAL APPLICATIONS

The amateur is often tempted to use a species of "Old English" with the misguided hope of achieving a "fancy" effect. He had better resist the temptation and give way to his creative impulse only when the occasion to use Old English is most appropriate. The occasion may include diplomas and certificates, testimonials and festive holidays which reflect a religious nature. This includes Christmas, Easter, Thanksgiving and perhaps St. Valentine's Day. Extensive use is made of Old English on greeting cards for these occasions as well as for posters, window displays and other merchandising media. Traditionally Old English is used on advertising which features Santa Claus and his bag of gifts. In testimonials and proclamation lettering, Old English is fine, but don't cheapen it by misusing it. What's more it does not rate best in legibility. Save this alphabet for the holidays.

ABCDEFGHIJKL
MNOPQRSTUV
WXYZ
abcdefghijklmnopqr
stuvwxyz
1234567890

ESSENTIAL CHARACTERISTICS

This is one of the newer and more popular display faces in use today. Its main points of identification are the slab-type tapered block serifs, the extended width of the letters in general and the clean and consistent relationship between the thick and thin elements. Other distinctive tell-tale characteristics are the hook-tapered lower terminal of the G, the ball terminal of the J, which seems snugly nestled close to the upright stroke, and the swash of the Q, which enters the counter of the letter and balances itself out of the circle.

ESSENTIAL CONSTRUCTION

Fortune is a standard type face and is readily available at most typographic houses. Its popularity also is attested to by the fact that it is procurable in other forms as well, such as paste-up or transfer letters, photo lettering, pre-cut paper and cardboards, wood and three-dimensional letters. If, however, it is to be hand lettered, it is not a difficult lettering style to render. All letters are constructed in rather wide form giving the feeling of each fitting into a square. There are two main thicknesses and the relationship is kept without variations except, of course, around curves or circles. The mechanical structure of Fortune makes it an ideal letter for ruling pen and straightedge. The sturdy structure and absence of hairlines makes Fortune a practical choice for cut-out letters made of heavier board or wood. Similarly it is structurally fitted for metal.

ESSENTIAL APPLICATIONS

Fortune is extensively used in advertising appearing in newspapers, magazines and TV titles. It also is a favorite type face among typographers for books, especially for title pages, chapter headings and for book jackets. Because of the sturdy structural elements, Fortune Bold reproduces well in any size and in either direct or reverse printing. There are no thin elements to break down in the plates or in the actual printing. Because it is comparatively fast and easy to cut, this is a letter style you'll want to use for silk screen too!

ABCDEFGHIJKL
MNOPQRSTUV
WXYZ&
1234567890

■ ESSENTIAL CHARACTERISTICS

The special character of Forum is due to many subtleties, not at first obvious. Observe the slightly inward curve of the bottom of the serifs and the transitional blend which unites them to the main stems. Note the lower lobe of the B which extends beyond the upper lobe and is accented on a diagonal axis. This diagonal axis accent is one of the identifying characteristics of this classic letter form. You'll see it very clearly in the position of the weight of the D, O, Q, etc. Other identifying elements of Forum are the open lobe of the P, the bottom weighted accent of the swash of the R, the curved branch-like effect of the upper part of the Y, and the overlapping V elements of the W. Note too that the J seems unfinished since it appears to be devoid of a finishing terminal.

The numbers too are easily spotted, since they have retained the old classic style evident in the fact that they do not all line up top and bottom. For example, the 3 dips below the line; the 6 rises slightly above. Note too the extremely low cross strokes of the 4 and the diagonal cross stroke of the 5.

■ ESSENTIAL CONSTRUCTION

Forum looks as if it were cut in stone with chisel and mallet, as in the days of Roman artist stone cutters who evolved our basic standard alphabets we know today. The origin of the serif stems from these early artists of the chisel and represents a finishing touch for the start and ending of a stroke. The same need for serif endings pertains to calligraphers of pen and ink. Forum shown here is closely allied to its calligraphic past. The most natural tool for rendering this type face is a broad-nibbed pen or a well chiseled short-haired showcard brush. The best results are achieved if the nature of the tool dictates the formation of strokes, and in this case the diagonal axis of the accented stroke weights will best be achieved by using the pen or brush as a calligraphic tool. In the hands of a true calligraphic artist, this alphabet can be achieved effortlessly, with a minimum of strokes and artificial refinements.

■ ESSENTIAL APPLICATIONS

The classic structure of this alphabet suggests uses with a classic theme. There is something enduring about this alphabet, an impression which it reflects when used on window and over-the-door signs for banking establishments, institutions of learning, libraries, financial houses, etc.

Forum was born as an incised letter in stone, and is still for that purpose for architectural lettering on public and civic institutions. Its use however is easily extended to all media of paint, ink and structural material. It is a practical letter to design for three-dimensional materials of wood, metal and board.

The wide use of Forum extends to monuments and pedestal inscription, plaques, diplomas and certificates, books dealing with the arts and classics. This alphabet will be a good addition to your repertoire, to counter balance the growing number of current favorites whose lifeline is not as enduring.

ABCDEFGHIJKLMNO
PQRSTUVWXYZ
&!?$1234567890
abcdefghijklmnopqrstuvwxyz

ESSENTIAL CHARACTERISTICS

Some conservative typographers would easily vote this alphabet as the No. 1 display face. It's almost an adage in the trade to say, "when in doubt, use Franklin Gothic." This favorite among type faces is, as its name indicates, in the Gothic family. A closer view will reveal that it is not really a strictly one-thickness letter. This subtle variation in thickness is discernible in such letters as the A, the diagonal in the K, accented left stroke of the U, etc. The variation in thickness is much more marked in the lower case than in the caps. In fact, the lower case departs so much from the Gothic one-thickness structure that all the letters containing round elements such as the a, b, c, d, e, etc. are decidedly accented letter forms tending toward the Roman thick-and-thin family of alphabets.

ESSENTIAL CONSTRUCTION

Because of the decided versatility of Franklin Gothic, there are many ways to produce it; the ways varying with the ultimate use to which this style is put. For instance, for careful rendering on a poster or book jacket design suitable for reproduction, the letters are drawn out with care and followed up with inking by ruling pen. If Franklin Gothic is to serve as the heading of a hand-lettered display or exhibition panel, a showcard brush carefully guided by the deft hand of an expert will be the proper tool. For truck lettering a flat ferrule brush will produce the flick-sharp corners with sufficient crispness to give the desired effect. The particular version of this alphabet is one in the family which contains such variations as Franklin Gothic Wide, Franklin Gothic Condensed, Franklin Gothic Extra Condensed, Franklin Gothic Italic.

ESSENTIAL APPLICATIONS

If you were to be limited to half a dozen alphabets to work with, and no more, chances are Franklin Gothic would be in the lot. This is due to the fact that it is an alphabet for reproduction, showcards, sign painting, truck lettering, exhibit panels, silk screen reproduction, etc. Why is it so versatile? It is easy to read and not difficult to construct. It is a stable and conservative style, yet modern as a jet airliner. It prints well with any process and on practically any stock. It reproduces well direct (black on white) or reverse (white on black). Gosh, all this praise may indicate that I am personally prejudiced in its favor. I am, and so are thousands of others who prefer something of lasting beauty and serviceability.

ABCDEFGHIJKL
MNOPQRSTUVWXYZ
&!?:$1234567890
abcdefghijklmnopqrstuvwxyz

■ ESSENTIAL CHARACTERISTICS

This basic type face has an individuality all its own, which makes it distinctive; easy to recognize, easy to construct, easy to read.

Although at first glance Franklin Gothic Extra Condensed seems like a one-thickness letter, closer inspection will reveal slight modifications, as in the thinner diagonals in the K, M, V, W. Note that the outer curved elements of letters such as the C, G, O and Q have been retained, although these curves are usually completely flattened in an extra condensed letter form. However, the flattened curves are evident in the inside areas of these letters (C, G, O, etc.). Other identifying letters are the G, with the appendage on the lower right side; the fore-shortened center stroke of the E and F; the fore-shortened loop of the J which does not go below the base line.

■ ESSENTIAL CONSTRUCTION

In constructing this letter, think of it as a one-thickness letter first and then make the variations in thickness referred to under "Essential Characteristics." All the inside areas are of approximately the same width, such as the C, D, G, O, etc. The letters occupy essentially the same structural width, with the exception of the J, M and W.

Tools such as the T-square, celluloid triangle and ruling pen are useful in the mechanical construction of this alphabet. Once a template is made conforming to the structure of a letter such as the H, it forms the basis for mechanical measurement of most of the alphabet. The width of the H and the cross bar establish the basic structural pattern. The use of freehand brushes for the round elements will avoid an over-mechanical look.

■ ESSENTIAL APPLICATIONS

This Extra Condensed version of Franklin Gothic should be among the alphabets in your repertoire. It is one of the most basic styles in display and advertising type faces. Its simplicity makes it a logical choice for posters, displays, exhibition panels, and it's equally serviceable cut out in wood, metal, plastic, beaver board, etc. Film cutters prefer this alphabet because it can be done quickly. It is devoid of delicate variations in thickness, has no appendages or serifs to bother about, reproduces well in reverse and direct printing.

Keep this alphabet handy. You'll see how often it will be the answer to your search for an appropriate alphabet for ever so many uses.

Franklin Gothic Extra Condensed

ABCDEFGHIJKL
MNOPQRSTUV
WXYZ
&!?$1234567890
abcdefghijklmnopqrstuvwxyz

Franklin Gothic Wide

■ ESSENTIAL CHARACTERISTICS

The present trend in extended type design seems undiminishing. Pick up and examine any of the smartly designed magazine and newspaper ads and you will see the popularity of wide or extended types in use today. Franklin Gothic Wide is a thick and thin letter which seems to be made to fit into a square. The letters in most cases are as wide as they are high, the I and J being the principal exceptions. The main identifying earmarks are the slight variations in thickness of stroke between the thick and thin elements and the general squatty proportions of the letters. Other identifying marks are the short center cross bar of the E and F, the lower-than-center cross stroke of the G and the short appendage at the bottom of it and the tail swash of the Q.

The lower case also is extended but is decidedly thick and thin in very noticeable variations, as is evidenced in the lobe of the d, the cross stroke of the e, etc. Notice too, that both ascenders and descenders are very short.

■ ESSENTIAL CONSTRUCTION

This is a comparatively easy letter to construct. Devoid of frills and typographical excesses, Franklin Gothic Wide is basic in structure. The caps are constructed along the block type with two essential thicknesses. The lower case does not quite follow the basic block form. The thicknesses vary between letters. Essentially this is a reproduction alphabet intended for ruling pen and India ink. If the letter is to be indicated in "comp" form in pencil, a chisel edge carpenter's pencil will produce the desired effect. In the deft hands of a good comp letterer, each stroke can be rendered in one movement of the pencil, provided the width of the chisel edge is the full width of the stroke.

■ ESSENTIAL APPLICATIONS

Franklin Gothic Wide is an ideal newspaper layout style. It is versatile in the effect it is intended to produce. It would be as appropriate to use on an ad selling real estate as it would be advertising the services of an industrial designer. Because of its impact value and excellent legibility, it is also a favorite for posters, displays and billboards. It reproduces well for direct and reverse printing and is acceptable to any printing method.

ABCDEFGHIJKLMNOP
QRSTUVWXYZ&
1234567890$
abcdefghijklmnopqrstuvwxyz

ESSENTIAL CHARACTERISTICS

This is unmistakably a pen letter style. It does not rank with the esoteric calligraphic scripts, but rather conforms to the widely known "Speedball" pen lettering styles seen on high school diplomas and certificates.

The capital letters are a blend of informal lettering and a kind of vertical handwriting. It may be classified as an upright text script. While the upper case letters are made round and full bodied, the lower case letters are consistently simpler in structure. And aesthetically the better for it! The structure of the terminals of all strokes, straight as well as round suggests a set angle of the brush. This characteristic is discernible in both upper and lower case.

Freehand, as a type face was cast by the American Typefoundry, and has its counterpart (with some variations) in other "pen lettered" type faces, such as Heritage and Vernon. The Freehand type shown here comes in sizes ranging from 6 to 36 points.

ESSENTIAL CONSTRUCTION

The Freehand alphabet shown on the opposite page was designed as a pen lettered alphabet, and is intended to be used by those who do not think in terms of metal type, but rather of lettering suitable for hand lettering.

A broad-nibbed lettering pen is the natural tool for this alphabet. Each movement of the pen should yield a stroke with a measure of authority and finality. No stroke should be repeated or be artificially "doctored up." The pen should not be rotated in the manner of a brush, nor should the modulated weight of the stroke from thin to thick be caused by variation of pressure on the pen. The change-over stroke from one thickness to another should be the result of the angle at which the pen is held rather than a flexibility in pressure. The nib of the pen when guided sideways will produce a fine hairline while the same nib when held broadside will yield a brush stroke the thickness of which will be determined by the angle at which the nib contacts the paper.

ESSENTIAL APPLICATIONS

This is a very practical alternate to the more ornate Old English (shown on page 82). It is quicker to do, has greater legibility and can be used on more occasions than Old English. While the traditional Old English alphabets are inextricably associated with Christmas and Easter holidays, Freehand has a far greater range of application. It too is suitable for holiday greetings, diplomas and the like; but in addition, it is equally serviceable as a fast lettering style for showcards for any occasion—and for any season. As a showcard letter, it has the advantage of speed on the job where time is of the essence. A good pen letterer can practically "write" the copy using this style, and with speed (which comes with practice) develop a natural rhythm which is unattainable when each letter is done in slow tempo.

ABCDEFGHI
JKLMNOPQR
STUVWXYZ

■ ESSENTIAL CHARACTERISTICS

It's easy to render this type face. Think of it as the letter with a flower in its lapel. Actually the decorative motif which ornaments the center of each heavy stroke is an oval-shaped embellishment which gives Fry's Ornamented a rich sparkle distinguishing it from the more plebeian members of the family of alphabets.

A less lyrical analysis of this letter form will show it to conform basically to the traditional Roman letter structure, with the usual relationship of thick and thin strokes. The serifs are somewhat spurred and consistent throughout. The heavy strokes however are really a combination of two parallel strokes; one thin, the other bold. The horizontal cross bars of the strokes have a single line thickness, which branches out in triangular shapes serving as open-faced serif endings, very obvious in the E and F. Note too, the rather wide proportion of both of these letters and the inordinate width of the W, which seems like two V's strung together.

■ ESSENTIAL CONSTRUCTION

As in all open-faced or inline letter forms, it is best to temporarily disregard these embellishments in laying out the work in the initial stages, preparatory to the actual finished lettering. Assume for the time being, that you are laying out solid stroke letters, so that your sense of spacing will not be confused with nonessential elements. When this is accomplished to your satisfaction, it is a simple matter to add the inline, ornamentation and doodads which add special character to the particular letter style. It will prove to be fun for you to experiment with using an additional color for the embellishments. This added color may be applied as a transparent wash within the white space of the double thick strokes or the oval area. This should help to give additional sparkle to an already ornamented letter form and may be just the "touch" that will make your customer realize that you are more than a letterer—you are an artist!

■ ESSENTIAL APPLICATIONS

The dainty structure of this letter form must be given dainty jobs to do. You should not make a fawn pull a heavy plow. Fry's Ornamented is suited for merchandising which deals with flowers, perfumes, dinnerware, expensive glass and other things which appeal to the carriage trade. The very thin cross strokes would make it impractical for cut-out letters of board, wood or metal. Care too, must be exercised in printing, as improper makeready or ink mixture will blur the white inlines; a caution equally important whether the reproduction be silk screen, letterpress or gravure. For silk screen reproduction intended for hand-cut stencil, this would not be a good letter to choose since it would entail a tremendous amount of cutting time. But then with a hundred type faces to choose from, why not select the face best suited for the product and the process by which it will be executed. Save Fry's Ornamented in your file marked "special occasions." When the time comes, reach for that drawer.

FRY'S ORNAMENTED

ABCDEFGHIJKLMN
OPQRSTUVWXYZ
&.?$1234567890
abcdefghijklmnopqrstuvwxyz

ESSENTIAL CHARACTERISTICS

The type face shown on the opposite page has such marked characteristics that it is completely different from all other alphabets. Each letter of the alphabet carries the common identification; a blocky disjointed grouping of the two or three elements which give identity to the letter. In essence, the main shapes are a rectangle, a triangle and a half circle. There is, as you will note, a complete absence of connecting hairlines, thus producing a sort of stencil effect. The general impression is one of toy-like blocks of wood placed together in such a way that each letter is recognized in its minimum structure. There is no other alphabet quite like Futura Black, and you'll be able to spot it immediately, now that you know its real name.

ESSENTIAL CONSTRUCTION

Futura Black is one of the easiest letters to construct, so easy, in fact, that children in grade school learn it for use on posters, monograms, etc. For professional standards, a certain degree of care is necessary to obtain the desired effect. All strokes are carefully measured off in pencil before ruling pen and fill-in brush are used. The round elements of the O and Q, which resemble two hemispheres, are produced with compasses. Most letters fit into a compact rectangle, as is especially noticeable in letters such as H, N and Z. As stated above, this alphabet is easy to do. However, it takes a skilled professional to do it well. When done well, Futura Black is in a class by itself.

ESSENTIAL APPLICATIONS

Although this style was designed more than 20 years ago, there is something ever modern about it, which suggests a timeless elegance. Years ago, as today, you would see this style used over the store fronts of the finest shops on the boulevards of Paris, London, Rome and on New York's Fifth Avenue. Futura Black is an exclusive alphabet, reserved for expensive exhibits, very modern delivery truck lettering and store signs selling flowers, lingerie, cosmetics and other fineries. It is also an excellent style for letterheads, especially when used in combination with a light script, which will contrast with it harmoniously. For best effect, only one line of Futura Black should be used. This alphabet does not rank high in legibility and should not be made to do anything more than to create an impact or impression. Let the simpler letter styles which can be used with it carry the burden of the copy.

ABCDEFGHIJKLMN
OPQRSTUVWXYZ
&!?$1234567890
abcdefghijklmnopqrstuvwxyz

■ ESSENTIAL CHARACTERISTICS

The Futura family is an alphabet of many faces—and there is very little family resemblance between them. Look at the Futura Bold Italic shown here and the Futuras shown on page 99 and page 103 for comparison and contrast. The alphabet shown here is an Italic variation of the standard Futura Bold where the letters are upright rather than slanted as in the Italic. The difference between them is only in the angle. Therefore this example will in effect be sufficient for both variations.

In essence both variations have this in common. One thickness (in appearance, not in actual measurement) for the upper case and a marked variation in thickness for the lower case. The O in the standard Futura is a compass circle, while in the Italic it is an "oblique" circle. And, of course, both variations are sans serif.

■ ESSENTIAL CONSTRUCTION

Think of Futura Bold Italic (and its companion standard upright Roman type) as a one-thickness letter throughout the upper case and construct it with that in mind. Then, add the subtle refinements in thickness. The cross bar in the A is made a shade thinner. That also applies to the cross bars of the E, F, H, etc. The sides of the M are flayed and are not the same angle as the upright of the L or N.

In doing the lower case it is necessary to show greater variation in thickness due to the fact that the counters of the letters would get too filled up and black looking if all strokes were kept uniformly heavy. You can see the reason for this clearly in the a, e, g, p, q, etc. The ascenders are made somewhat longer than the descenders and with the exception of the f all ascenders (and descenders) are square finished strokes without hooks, curlecues or other structural kerns. Note the j for instance, which is merely an elongated i.

■ ESSENTIAL APPLICATIONS

Futura Bold and Futura Bold Italic are variations of Futura and Futura Italic, the difference being only one of thickness. The Futura Bold type faces are among the most popular display faces for newspaper and magazine advertising headings, feature copy for direct mail advertisings, as well as for posters, window displays, 24 sheet posters, car cards, and nearly every promotional job in print. Because of its utter simplicity, excellent legibility, it has been the "stand by" of poster artists, typographers, layout artists, and book jacket designers. In addition to its visual quality this alphabet (and its variants) have excellent printing qualities. The printer welcomes these type faces because there are no delicate parts to break down or close up in printing.

FUTURA Bold Italic

ABCDEFGHIJKLMNOPQR
STUVWXYZ
abcdefghijklmno
pqrstuvwxyz
1234567890

ESSENTIAL CHARACTERISTICS

Where you want compactness and power, there are not many alphabets which will answer that need more than Futura Display.

It is massive in weight and although the strokes appear to be of uniform thickness, closer inspection will reveal a variation in weight, clearly evident in the M and N. The lobes of the B as well as all other lobes have been flattened so that both sides of the letter fit into a vertical rectangle. The only roundness retained is in the "elbow curves" or corners. The counters are rectangular and of uniform thickness. Special note should be made of the K which has no diagonals; the lower case type of M and N; the peculiar hook extension of the Q. The V is unusual in that the strokes change direction as they approach a joining at the bottom; the W is a combination of two U's and the X is of an hourglass format.

The lower case reflects the same massive compactness in structure and appearance. Note that even the dots over the i and j are square and close to the main body of the letter. The m, n, o, s, u, v, w, x and z are identical with the upper case, but of smaller format.

ESSENTIAL CONSTRUCTION

With practice, this alphabet can be rendered very quickly with a flat well chiseled brush—guided by the hand of an accomplished letterer.

Futura Display, because of its uniformity in the space that each letter occupies (with some variations) is a simple letter to space. For reproduction work, the letters must be laid out carefully in pencil. The straight lines may be ruled in with ruling pen, but the curved elements require the use of a brush. Even some of the strokes which appear at first glance to be perfectly straight, will upon closer inspection be found to have a very slight freehand curve, as in the upright element of the E, G and Y. You will also note a faint semblance of a serif in the first upright stroke of the M, N and ending strokes of the K and R.

Silk screen film cutters will welcome this heavy type face because it is comparatively easy to cut out. Most strokes can be cut with aid of a straight-edge. Slight accidental unevenness in thickness will not be too easily discernible because the alphabet's heavy weight will absorb any small irregularities. The elbow corners however must be kept uniform throughout.

ESSENTIAL APPLICATIONS

Put this in your top drawer marked "alphabets for poster work." Futura Display is one of the most reliable work horses for lettering for posters and display work. It has an audacious impact, is very legible, is rendered without too many demands on technique or time. It is an ideal letter for industrial exhibits either in painted or three-dimensional cut-out form. The caps are more frequently used than the lower case and both are available as printer's type in point sizes 14 to 84. If larger size is needed, this type face lends itself very well to photographic enlargement without loss of character, detail or sharpness. It is an excellent choice for billboards and poster advertising—yet looks equally good for headlines in newspaper and magazine ads.

Futura Display

ABCDEFGHIJKLM
NOPQRSTUVWXYZ
1234567890$?!&

ESSENTIAL CHARACTERISTICS

This is a standard type face designed for commercial use, but it resembles a playful attempt at doodling with letters of the alphabet. This style of letter reminds one of the kind embroidered on a silk shirt, engraved on cuff links or printed on personal letterheads. The playful doodly-type thin strokes are very evident in the graceful cross bar of the A, the lobes of the B, and the swirl curves of many of the other letters. Basically Gallia is a thick and thin Roman letter, with strong contrast between the thin and thick strokes. The main strokes are heavy, made to appear airy with the center inline within the thick strokes. The serifs are thin hairlines which evolve into the thin element of the letter form, especially evident in the lower strokes of the E, L, etc. The delicate thin lines are as graceful as the hairspring of a watch, but closer inspection will show that they are weighted in the center.

ESSENTIAL CONSTRUCTION

The best way to render this ornate style is to first lay it out without the heavy center inline, as if the heavy strokes were solid. When this procedure is followed, there will be a minimum of visual confusion in spacing or pre-occupation with refinements. Indeed, it would be advisable to render this (with brush or pen and ink) as outlined skeleton letters devoid of the solid heavy stroke. With the outline completely finished, it is a simple matter to add a heavy stroke centered within the outline. Very colorful effects can be achieved when these center strokes are made in a different color, for a flowery and sparkling appearance. Another way to achieve an interesting effect is to treat the heavy strokes as solids, with a center stroke painted in a different color. Gallia is a "fun letter" you'll enjoy working with.

ESSENTIAL APPLICATIONS

In addition to the personalized uses mentioned before (letterheads, initials or personal items, etc.) Gallia has very definite commercial applications. Showcards and posters advertising flowers, perfumes, women's wear and other items which express femininity and gaiety could be made attractive and appropriate for the occasion. Gallia in any of its varied forms is a good style to try on merchandising of silverware, furs, menu covers, optical goods, etc. It is a standard printers' face, one which will afford you many tempting opportunities for personal variations with brush or pen.

ABCDEFGHIJKLMN
OPQRSTUVWXYZ
&!?.$1234567890
abcdefghijklmnopqrstuvwxyz

ESSENTIAL CHARACTERISTICS

A classic Roman thick-and-thin alphabet, Garamond Bold is more popular now than ever before. Exotic "fad" type faces capture the popular fancy but have a short life; the basics survive to be newly appreciated by generations of discriminate tastes.

Outstanding identifications of Garamond Bold are: the pointed apex of the A which rises above the top guide lines; the swash of the J which dips way below the rest of the letters and ends in an oblique curve; the slightly slanted side strokes of the M; the odd shape lobe of the P; the virile tail-like swash of the Q; the construction of the W which can be best remembered as two V's overlapping in the center. The serifs are substantial and you will note that they blend and "melt" into the main elements of the letter form, especially noticeable in letters like the B and D.

ESSENTIAL CONSTRUCTION

If a brush facsimile of this type face is to be rendered, a flooded brush is recommended. A fully charged flooded brush will naturally produce the full serifs and the blended transitions of curve and straight line referred to in the "Essential Characteristics." This is a built-up letter and is suitable for pen work as well as for showcard brush. For a meticulous rendering suitable for reproduction, Garamond can be partially done with a ruling pen, but compasses cannot be adequately used here because all curves are natural and not geometrically symmetrical, with the possible exception of the outside curvatures of the O and Q. All other curves and lobes require the skill and feeling of the hand lettering artist.

ESSENTIAL APPLICATIONS

The classic nature of this type face suggests a conservative setting and function. This is no letter to employ for slap-dash effects. Garamond (in the upper case) is in its proper elements on title pages and chapter headings for books. Since it is both a display as well as a "body copy" type face, the lower case is a serviceable type for complete typographic duty on books, magazines and advertising copy. The lower case however, is seldom used for signs and displays. While the upper case alphabet is occasionally seen on oilcloth signs and window cards, it is much more appropriate for fine work for office doors and on gold leaf glass work. It is shared by sign painters and typographers.

ABCDEFGHIJKLMN
OPQRSTUVWXYZ
&.,!?:;$1234567890
abcdefghijklmnopqrstuvwxyz

■ ESSENTIAL CHARACTERISTICS

This is a strange—even awkward-looking type face when each letter is seen as an isolated unit; but paradoxically enough, the apparent lack of good form and consistency seems to vanish when the letters are united to form words. Indeed it is difficult to analyze all the letters of this alphabet because each one has its own peculiarity. This is one alphabet which refuses to be judged individually, but let's attempt it anyway with the hope of arriving at some generalization.

The slant of the lettering varies, as you will see when you keep the I next to the J. The relative width of the letters vary. Look at the wide spread of the flayed M and the condensed oval-shaped O.

These variables and inconsistencies apply even more to the lower case. Compare the thickness of the upright stroke of the l and that of the p. Note the drooping swash tail of the z; the tucked-in bulging right element of the h, etc.

The need for "togetherness" that each letter of the alphabet has for the other is typical of some of the letters of the best calligraphic alphabets which depend upon each other for tone and cohesion.

■ ESSENTIAL CONSTRUCTION

Although this is a standard type face, it presents interesting possibilities to the hand lettering artist as a model for personal variations. The lower case especially may serve as a source of inspiration for a fast showcard lettering style which lends itself equally well for brush or pen. Of course the speed with which this style is rendered will depend not only upon the skill of the letterer but the nature of the occasion. For posters intended for reproduction, the letters will have to be carefully pencilled in, outlined in fine brush and filled in. For routine showcard lettering, an effective facsimile can be achieved with a one-stroke technique, using a lettering pen or a well-paleted chisel edge showcard brush.

In this type face, the lower case letters are used far more frequently than the caps. When using the lower case, it is best to space the letters very close together, a rule that holds good for most italic lettering.

■ ESSENTIAL APPLICATIONS

The Garamond family of type faces has a tradition that reaches back to the 16th century French typographer, Claude Garamond, from whom the name is derived. Since then the design has undergone many revisions yet retaining the same character identity.

Garamond type faces are not well known among sign writers and poster artists who are not involved with printing. This is a printers' type face traditionally and is usually used for book publication, both as a display face and body copy. It is also extremely useful for all types of printing needs, from a simple business card to an annual report. It ranks with Caslon as one of the basic printing types of the graphic arts field.

Garamond is readily available as a standard form for photo lettering units, and can be specified from most composing rooms with the assurance that it is always in stock.

Garamond Italic

ABCDEFGHI
JKLMNOPQR
STUVWXYZ
abcdefghijklmnopqr
stuvwxyz

ESSENTIAL CHARACTERISTICS

Like all good art, this alphabet has the imprint of the tool which helped to shape it. In this case, the tool is a broad-nibbed or chiseled instrument such as a flat pen or brush. The beautifully accented variations in the strokes give Gloria Bold a calligraphic appearance. This type is based on the flourishes of a fanciful yet firm specimen of penmanship. Among the letters which lend themselves especially to identification are the A, W, X, each with a rather singular cross stroke which is strikingly different. In the lower case, the ascenders are tall, the descenders rather short and the peculiar dots of the i and j lend a mark of distinction to this alphabet.

ESSENTIAL CONSTRUCTION

While Gloria Bold is obtainable as a type face, it is one which you may enjoy doing by hand using either a flat brush or broad-nibbed lettering pen. It has a distinct calligraphic profile which makes it a personal type, one you should not try to copy slavishly. Study the slant, the play of thick and thin, the turn of the flourishes—and then proceed to be yourself and create your own version as your creative impulses and the natural flow of the tool work with you. Of course where you employ this type face as a type indication for the typographer, then a more accurate facsimile of Gloria Bold will be in order.

ESSENTIAL APPLICATIONS

Although Gloria Bold is a display face, it can be used on limited copy on department store showcards. As such it is appropriate for display work dealing with feminine apparel, cosmetics and luxury items. Like Old English and other calligraphic alphabets, Gloria Bold should not be used in caps alone. The capitals are reserved for decorative initials or the first letter in a word. Because of this calligraphic quality, this is a good type to substitute for Old English—it is simpler to do and easier to read. As such it can be used on certificates, diplomas and the like. Gloria Bold should prove to be an interesting addition to your growing stockpile of alphabets.

ABCDEFGHIJKL
MNOPQRST
UVWXYZ
&!?$1234567890

ESSENTIAL CHARACTERISTICS

In the days when there was more gold in "them there hills" than there was in the U.S. mint, this alphabet (or something similar to it) was used on the painted signs over the saloon, prospector's office, jail house and the general store of our western towns. That "golden era" of our history has disappeared but this alphabet so popular in those days has been currently revived along with a large range of other "old face" alphabets. The outstanding points of identification of the Gold Rush alphabet shown here are: a Roman letter structure with very little contrast between thick and thin; square, long "slab" serifs of the same thickness as the cross bars; an oblique shadow outline, placed to the right and lower side of the letter. Minor points of recognition are: the point ending of the V-shaped center element of the M; no serifs on the crossbars of the E and F; the W formed by two V's united by three slab serifs across the top. The numerals are consistent with the caps with the exception of the peculiar open top structure of the 4 and the bend of the top line of the 5.

ESSENTIAL CONSTRUCTION

A good rule to follow in rendering this letter style by hand, is to lay out the work as if there were no shadow outline. The shadows introduced at the time of the initial layout would only confuse your sense of spacing and you would get so entangled with these, that you would not see the forest on account of the trees. After the layout is complete and passes your critical inspection, then add the shadow elements. In fact, for most routine sign and showcard work, brush in the lettering completely adding the shadows last, as an operation by itself. Instead of the line shadow structure, you may use another color, preferably one that is subdued and will not compete with the color used on the main structure of the letter. The shadow will give the lettering a three-dimensional quality especially if it is treated as a solid color area.

ESSENTIAL APPLICATIONS

For masculine appeal, Gold Rush rates high among those of us who have a hankering for the good old days. Gold Rush is a man's letter not intended for women folk and sissies. It is a natural for men's cosmetics, tobacco, menu covers for popular steak houses, etc. Its use can of course be extended to things archaic, antique or historic. Used with discretion, Gold Rush can be combined with a modern type face, thus making one letter style the contrasting foil for its companion.

Gold Rush is a standard type face but it was never designed with a lower case. This gives you a chance to design an appropriate lower case of your own to match the caps.

ABCDEFGHIJKLM
NOPQRSTUVWXYZ
&!?.$1234567890
abcdefghijklm
nopqrstuvwxyz

■ ESSENTIAL CHARACTERISTICS

Here is easily one of the most classic alphabets in our repertoire of type faces. The late Frederic Goudy who designed it, found his inspiration for it in the inscription frieze of the Trojan column and modified what he saw to a perfection which reflects current standards of legibility combined with classic beauty. This type face is one of a series of Goudy designs all distinguished by careful symmetry and well-modulated proportion. It is a Roman alphabet at its best. Some identifying letters are the M with the diagonal center structure extending on both sides, the tilted O and Q, the crossing V's which comprise the letter W, the swash tail of the Q which does not enter the main body, the below-the-guide-line hook of the J.

■ ESSENTIAL CONSTRUCTION

To do a Goudy alphabet justice, you really have to be a crackerjack letterer. The subtle tapering of the strokes cannot be duplicated with mechanical instruments and mere precision. To say you must "feel" this letter is no definite help in instruction in its rendering, but there is no formula that can be followed in this as there can be in a Kabel or other instrument-type of letter form. What will flow out of your brush will depend upon your experience as a letterer and your aesthetic sense of good design.

■ ESSENTIAL APPLICATIONS

The classic quality of Goudy Oldstyle suggests uses where restraint and permanency are important. This is a letter that can be cut in stone with assurance of its timeless reverence for what is eternally beautiful. Goudy is a good choice for books of classic appeal, inscriptions for banks, houses of worship and educational institutions. Commercially, it can be used for fine letterheads, office doors, high calibre display panels for exhibitions, etc. It is not a good choice for average posters or occasional signs that are viewed "on the run," but Goudy Oldstyle can be employed for store window cards for fine haberdashery and better department stores.

ABCDEFGHIJKLMNOPQRSTUVWXYZ

abcdefghijklmnopqrstuvwxyz?!

1234567890

■ ESSENTIAL CHARACTERISTICS

The alphabet shown here is called Goudy Text and is a standard type face familiar to most printers. Call it Goudy Text or what you like, it's still Old English to most people. There are hundreds of variations to Old English, but only an expert can tell the difference between one and another.

This alphabet is a creation of the great American type designer, Frederic Goudy. Note that the upper case has round elements and ornamental flourishes, while the lower case is elongated, vertical and devoid of frills or curves. The calligraphic pen quality of this type face is dramatically shown by the rather virile heavy main vertical strokes contrasted by the hairline thin and wispy serifs.

■ ESSENTIAL CONSTRUCTION

This is a favorite alphabet among amateur artists who like to do "fancy" lettering. But whether it is produced by amateur or professional, the pen is the standard tool for its construction. A flat broad-nibbed pen held slightly at an angle will yield to pressure to give the full weight to the accented strokes. The thin strokes are produced by using the side of the pen. For a more elaborate treatment, the thin lines may be added as a separate operation with a crow quill or other fine penpoint. The possibilities and natural temptations for improvisations are so great, that unless an exact copy of this type is absolutely mandatory, it would serve no purpose here to call particular attention to every twist, turn or flourish.

■ ESSENTIAL APPLICATIONS

For things or events associated with the religious, festive or archaic, this type of alphabet is traditionally acceptable. It is the style of diplomas and testimonial prose. Commercially, Old English type faces have been universally adopted for the expression of season's greetings on cards, posters and displays for Easter, Christmas and other happy holidays. It is inappropriate for routine signs because it is not too readily readable where urgency is important. No Old English type should be used only in caps. The caps being highly ornamental should be used for initials only.

ABCDEFGHIJKLMNOP
QRSTUVWXYZ
&?!$1234567890
abcdefghijklmnopqrstuvwxyz

ESSENTIAL CHARACTERISTICS

The calligraphic quality of Grayda is one of its most noted characteristics. Essentially it is a thick and thin alphabet, based on handwriting with accented tones and interesting interplay of light and dark. The capital letters are free and have extending elements which dip below the guide lines, as you will note in the A, C, G, etc. The lower case is somewhat more disciplined and consistent. The most noteworthy common trait which all letters in the lower case have are the heavy accents on the tops and bottoms of each letter which swell several times the thickness of the thin elements.

ESSENTIAL CONSTRUCTION

Grayda lends itself to flat pen or well-paleted brush work. In either case the instrument is handled like a carpenter's pencil where the tool is not twirled between the fingers. Instead, the tool is held chisel-like in one direction with the thin elements constructed by the narrow part of the edge of the chisel. While Grayda is a well known and definite typeface, a personalized and free interpretation can be achieved by retaining the general effect without slavishly copying all the flourishes. After some practice a Grayda-like alphabet can be developed which can be as personal as your own handwriting. This is especially true of the caps which are embellished with flourishes and swirls.

ESSENTIAL APPLICATIONS

As a type face, Grayda is widely used for ads in newspapers and magazines. It is a "must" in the repertoire of any modern typographic house. In its hand-lettered version carefully rendered for re-production, Grayda, (or its equivalent) is appropriate for book jackets, letterheads, labels, etc. Rendered freely and casually this type face may be the inspiration for a hand-lettered fast script for pen lettered counter displays and cards. One word of advice. Grayda is a *display* face and too much of it on a job may be difficult to read or sustain interest. Use Grayda for captions, an occasional few lines, but don't let it carry the complete burden of a message.

ABCDEFGHIJKLM
NOPQRSTUVWXYZ
&.!?$1234567890
abcdefghijklmnopqrstuvwxyz

ESSENTIAL CHARACTERISTICS

This is one of the more modern type faces. How it got its name we don't know. It certainly does not seem to deserve a pug ugly name such as Grotesque. Be that as it may, here are some of the outstanding points of identification. First, it belongs to the Gothic family, judging from the fact that the strokes are not contrasting in thickness and that there are no serifs. Second, though the structure of each letter is condensed somewhat (similar to the Gothic commonly known as Gaspipe), it differs from that species in that there are slight curves to the round elements, as evident in the B, C, D, etc. There is a definite attempt to achieve uniformity in the widths of the letters with the obvious exception of the M and W, which are wider. The lower case is decidedly thick and thin as you will note when you study the a, b, d, e and practically every letter of the alphabet.

ESSENTIAL CONSTRUCTION

This type face is not a showcard writer's alphabet, simply because it would take too long to render it with speed. The precise nature of its structure demands an unhurried rendering, perhaps best achieved with ruling pen and ink. It might be said that Grotesque is a semi-reproduction letter. As such it is quite practical for exhibit and display panels, in which case the letters would first be accurately drawn out in pencil outline and then serve as a cutting layout for a paper frisket. This cut frisket would then be used as a stencil for airbrush or spray gun. With the evergrowing popularity of deluxe exhibition panels for the "big" shows at convention halls, it is not at all a rare occurrence for a silk screen stencil to be made for running off only one copy. This alphabet is a very popular choice for such assignments.

ESSENTIAL APPLICATIONS

We already have mentioned some uses for Grotesque. Other applications would fall in the categories of fine posters, titles for book covers, brochures, letterheads, credit titles for the movie and television industries and special truck lettering jobs which require a stately and conservative appearance.

ABCDEFGHIJKLMN
OPQRSTUVWXYZ
1234567890

■ ESSENTIAL CHARACTERISTICS

This alphabet is one of many type faces that form the group of condensed Gothics currently so popular. Most of them share the same essential characteristics—namely, all letters are more or less even in thickness, compact in appearance and solid in structure. This one, called Headline Gothic, has its own identity because of several characteristics which distinguish it from the other condensed Gothics. The J, M and Q are the identifying letters. You will note that the hook loop of the J reaches very high, coming up to the level of the lower part of the cross bar of the E, F, G, H, etc. The V-shaped center element of the M is short and comes to a point without reaching the bottom; the tail of the Q is clipped and does not go below the base line. Notice too, the W which is conventional and very much out of keeping with the M.

■ ESSENTIAL CONSTRUCTION

Although there are minute variations in thickness as in the round elements of the O, P, etc., Headline Gothic is almost mechanically uniform in weight of strokes. This means that a measured-off template can be used to mechanically assure even thickness throughout. In fact, the mechanical structure of the alphabet makes this an ideal one for use with drafting instruments, T-square, etc. The round elements, however, require freehand rendering as they do not form any part of a circle. This is a reproduction lettering style, usually rendered very meticulously. It is first pencilled-off in measured width, outlined with a ruling pen, with the round elements done with a fine lettering brush, and then filled in very carefully with India ink or paint. Watch out for spacing; the rule to follow is one which applies to all condensed mechanical style letter forms. Do not measure off mechanically the distances separating the letters. When a straight line of one letter faces another straight line of its neighboring letter, allow more space. Remember: It is not the *linear distance* between letters that makes for even spacing—it is the *area*—the visual volume of space which really is important.

■ ESSENTIAL APPLICATIONS

As the name indicates, this alphabet is primarily designed for headlines. This, however, is not to be implied to mean headlines only for newspapers. It is an excellent type face for all main copy headings—posters, displays, book jackets, exhibition panels, etc. It is an ideal letter to use because it is easy to construct, can be mechanically rendered, is easy to read and reproduces beautifully. What more can you want from a type face? Headline Gothic will be one of your favorites!

HEADLINE GOTHIC | **123**

ABCDEFGHIJKLMNOPQRSTUVWXYZ

abcdefghijklmnopqrstuvwxyz

&!?$1234567890

■ ESSENTIAL CHARACTERISTICS

Hellenic Wide reflects the current popularity of extended type faces. How long this vogue will continue is hard to predict. But that it's here is evident only too well when you turn the glossy pages of any newsstand magazines or examine closely typographic layouts produced by the best designers today. How to recognize the alphabet shown here? Easy! In addition to the extended width you'll note that this is strictly a one-thickness alphabet, including the serifs. And while we're on the serifs, note that they are squared off, are the same on either side of the main strokes and that they are extra long. Hellenic is a simple type face, consistent in basic structure and devoid of frills or unessentials.

■ ESSENTIAL CONSTRUCTION

This type face is an easy one to learn and render. Once you know that it is wide, one thickness and finished off with long square-trimmed serifs, you can easily render it without slavish or repeated copying. As explained above, the structure is consistent in thickness of stroke. Judgment enters in determining the relative proportion of height to width of the various letters. The round letters such as C, G and O are comparatively narrow. The structural requirements of each letter vary and accordingly the comparative width of each letter reflect those requirements.

■ ESSENTIAL APPLICATIONS

Although Hellenic and other extended type faces that we see around today are considered "modern," it seems to me that I have seen this type face shown here (or something very similar) on sides of railroad cars many yester-years ago. Hellenic is presently very popular for letterheads, labels, brochures as well as for newspaper and magazine advertisements. It is frequently seen on window and interior posters for department and specialty stores. A wide and extended type face such as Hellenic is employed most effectively when it is combined with a flourishy script or other frilly alphabet and arranged in such a fashion that the contrast between the alphabets is dramatically heightened.

ABCDEFGHIJKLMNOPQR
STUVWXYZ
&!?$1234567890

ESSENTIAL CHARACTERISTICS

This type face which comes only in upper case is a distinct letter form with an individuality all its own. It has decorative appeal and lends itself well to color variations which we will discuss later.

Foremost among its identifying characteristics is the inline which breaks up the mass of each heavy stroke. This inline is uniform in thickness throughout, and is mechanically centered within the main stroke. Another, though less discernible element, is the tiny spur-like serifs which appear like flicks on each stroke ending. Not noticeable at all at first glance, though contributing to the general appearance of Homewood, is the delicate tapering of all upright heavy strokes, as well as the thin horizontals. In the uprights, the heavier portion is at the top and tapers almost imperceptibly as it approaches down to the bottom.

The numbers, as shown here, though intended to follow the family pattern unfortunately do not come off as well, mainly because the thin lines vary in weight in the figures 2, 4 and 8. But this may be a personal criticism, not shared by you.

ESSENTIAL CONSTRUCTION

This alphabet is shown here as a black and white one-color letter form. As a type face it is intended to be printed in one color, but as a hand lettered alphabet you will have the opportunity to introduce an additional color in the thin inline which splits the heavy center stroke. This will add sparkle to any word without too much effort or time.

As in all decorative alphabets, it is best to lay out the words in solid form, omitting for the time being the inline. When the spacing seems right, then and only then, go back to the embellishments which give the letter its special character.

Since the strokes taper, it is not practical to use a T-square to facilitate mechanical accuracy. A ruling pen may be used, but each line will have to be judged individually and cannot be produced by using a triangle sliding along a fixed horizontal bar.

ESSENTIAL APPLICATIONS

All decorative alphabets must be used sparingly or confusion replaces ornamentation. Homewood can be a very effective and colorful headline alphabet. When used in moderation as a decorative motif and focal spot, it can dress up a massive copy panel done in a more basic style.

This alphabet is an especially good one to reserve for occasional use on showcards with a gay theme; showcards or displays intended for a flower show, art exhibits, department store white sales, etc. It would be out of place on the more pedestrian though more practical utilities such as signs for the butcher, shoe repair store or tailor shop.

This alphabet is also very appropriate as a type face selected for book jackets, title pages, brochure headline copy and annual reports and organizational souvenir journals.

ABCDEFGHIJKLMNOPQRSTUVWXYZ
ABCDEFGHIJKLMNOPQRSTUVWXYZ
1234567890
&?!$¢:;

ESSENTIAL CHARACTERISTICS

Huxley Vertical, shown here, has been a standard type face for many years and it is unique in many ways. Its outstanding points of identification are "tall, thin and condensed." It seems taller than it really measures, because of its thin structural elements grouped compactly. The narrow width is maintained consistently The B is as narrow as the C, D, G, etc. Exception to this can be found in the M and W which require a wider space. The cross bars are below center, a pattern characteristic evidenced in all letters which have a cross bar. Note too, the extension of the cross bar beyond the main letter area. The strokes are uniform in thickness throughout the alphabet. To retain the same area of compactness, the letters C, G, O, etc. which are traditionally made round, are flattened to an oblong area, with only the top and bottom elements retaining the roundness.

The numbers show a departure from the low slung patterns of cross bars which typified the caps. Here the height of the cross strokes vary from the low 2 to the high reaching center cross stroke of the 5.

Huxley does not come with a lower case.

ESSENTIAL CONSTRUCTION

This is essentially a reproduction technique letter intended for ruling pen and straightedge. As such, this would in effect be a one-stroke letter since the spread of the ruling pen can be fixed so that each stroke represents one stroke of the pen. The round elements may either be done mechanically with a compass or be carefully rendered using a freehand technique. An experienced letterer would in most cases prefer to attempt this freehand. Where mechanical precision is not of the essence, this type lends itself to a free interpretation, in which case a small size brush or pen would serve the purpose very well. When using this condensed type style, it is well to remember to leave a lot of space between the letters to avoid an astigmatic effect. Also it is a good thing to bear in mind to limit the use of this style to one line of lettering—preferably a long line running across the page in a fence pattern. This is all the more effective if the other copy is lettered in a conventional style. Contrast makes for interest and variety in any layout.

ESSENTIAL APPLICATIONS

The delicacy of line structure of Huxley Vertical naturally suggests subject matter of an ephemeral or feminine character. This field would include products relating to cosmetics, women's wear and fashion. Its proud and haughty appearance typifies luxury, elegance and refinement. Because of its thin uniformity, any variation in thickness due to uneven pressure of the brush or pen will show up glaringly. For stencil cuttings, this means extreme care in judging uniformity in thickness. Any variation will become conspicuous in the finished print. Huxley Vertical is an alphabet you will not often use, but when the occasion does arise, you'll be glad you will have another "special" alphabet at your disposal.

ABCDEFGHIJKLMN
OPQRSTUVWXYZ
&?¿!¡1234567890

ESSENTIAL CHARACTERISTICS

If you have seen this type face before, it may have been on the labels of bottles, banners for political campaigns or advertisements which were designed to appeal to the masculine taste. Jim Crow is one of a series of alphabets which the tide has brought in with the flow of old fashioned type faces so popular today.

The determining characteristics of Jim Crow are the horizontal cross lines which in a graduated way cover the face of the letter form. Another identifying element of Jim Crow is the shadow effect which is used more to add to the general ornamental effect than to create a three-dimensional illusion. Excessive ornamentation was the distinguishing note of identification for most type faces of years gone by.

ESSENTIAL CONSTRUCTION

Only a labor of love could prompt any letterer to attempt to duplicate this face, line for line. It is shown here with the thought that it may lead to improvisations and modifications to render it more practical to achieve the effect without an actual facsimile rendering. Upon analysis, it will be found that this type face is a sans serif block letter with the corners cut at an angle. To this has been added a shadow effect, created from the upper left hand source of light. The horizontal lines vary in spacing, to give the illusion of a blend.

In the good old days when such type faces were at the height of their popularity, the bill boards and placards were printed from wood type, cut and engraved by hand. This is a modified version of such a type. Jim Crow has no lower case.

ESSENTIAL APPLICATIONS

This type face is associated with a masculine market appeal which includes tobacco and cigars, hunting and fishing, old time music halls and circus and bazaars, etc. However, it is not confined to a man's world exclusively. Like all "revival" alphabets, Jim Crow can be effectively used to evoke a sense of nostalgia for the "Old World" and its association with our American heritage. This type face may be used effectively for outdoor poster and bulletin advertising antiques, silverware, rustic furniture, the theatre, etc.

As a printer's type, Jim Crow may be used as an alternate to others in its class, for book jackets and title pages for literature dealing with the phase of history relating to the Old West, political factions or the bizarre.

ABCDEFGHIJKL
MNOPQRSTUV
WXYZ

abcdefghijklmnopqrstuvwxyz
&!?$1234567890

ESSENTIAL CHARACTERISTICS

All letters of this type face are of the same thickness and all round elements are compass-shaped, including the S. The apex of the A comes to a point, slightly above the cap line. Also note that the bottom terminals of the A are diagonal. The same applies to the upper and lower stroke terminals of the E, L, X and Y. The W crosses in the center, the Y departs from conventional construction.

In the lower case, the identifying letters are the b with its circle lobe, the e with a diagonal cross bar and the g which is composed of two circular elements. Note too, the clipped terminal of the lower part of the j; the pointed juncture of the diagonals of the k. The letters p and q look as if they were viewing each other in a mirror. The lower case w does not follow the pattern of its big brother, the capital W.

ESSENTIAL CONSTRUCTION

The mechanical nature of this letter style suggests the use of a ruling pen, compass, transparent triangle and T-square. The accuracy of the engineer is more important here than the aesthetic perception of the lettering artist. Kabel letters must be carefully drawn in pencil, ruled in with pen and ink and then filled in. All letters with the exception of the S, are mechanically rendered. The S, for obvious reasons will require a little freehand styling.

ESSENTIAL APPLICATIONS

Because of its simplicity and great legibility, Kabel Bold, also known as Sans Serif Bold, is extremely versatile and has very wide applications. This includes posters, display boards, truck lettering, letterheads, business forms, etc. Due to the strong elements of each of the letters, this alphabet lends itself for use in cut-out form in wood, board or metal. Kabel Bold can also be used to good advantage on board and paper displays and reproduces well for all printing purposes.

ABCDEFGHIJKLMNOP
QRSTUVWXYZ
&.!?$1234567890
abcdefghijklmnopqrstuvwxyz

ESSENTIAL CHARACTERISTICS

Kaufmann Script, designed by an art director whose name it bears, is one of the most practical of the connecting script type faces. One thickness in weight throughout, Kaufmann is characterized by a freedom of swing and movement typical of the handwritten styles. The upper case possesses great flourish and dash, as you can note in the D, F, H, etc. The lower case is somewhat more disciplined, both in essential structure and angle. At first glance, the letters of the lower case seem to be linked together, but if you look closely you will see tiny gaps between each letter. This was one of the earliest type faces designed with links that could actually be butted together by the type setter so that lower case words could be formed to simulate handwriting.

ESSENTIAL CONSTRUCTION

The pen quality of Kaufmann Script suggests the tool, namely, the pen. However, an interesting effect can be achieved with a flooded brush. In a hand-lettered version of this type face, there are many inviting opportunities for individual variations, especially in the formation of the upper case letters. The loops and flourishes can be executed in accordance with the mood of the artist and the spirit of the occasion. Experimentation is invited in terms of thickness, from a fine pen line stroke to a strong bold stroke. The lower case demands greater discipline and consistency in the slant. It is this consistency of slant which makes for rhythm.

ESSENTIAL APPLICATIONS

This is one of the basic "connecting scripts" in the type setter's job case. Its practical nature is evident in the variety of uses of this type face. You can see Kaufmann Script on posters, book jackets, labels, letterheads, magazine and newspaper ads, movie and TV titles, etc. The style is popular and practical because it reproduces well in all sizes; there are no fine hairlines to break down in plate making and printing; and above all, it is an easy letter to read. The caps are used as initial letters leading into lower case copy. In the hand-lettered version, this is an excellent style to select for all assignments calling for elegance combined with restraint.

Kaufmann Script

ABCDEFGHI
JKLMNOPQR
STUVWXYZ

&!?$;:¡¿
abcdefghijklmnop
qrstuvwxyz

1234567890

■ ESSENTIAL CHARACTERISTICS

Latin Wide is a display alphabet with stocky triangular serifs which add additional width to an already expanded letter form. These serifs or "wedgies" help to line up the letters so that they carry the illusion of a continuous pattern of solidity. This is especially true when a complete line of words is viewed. The serifs are so stocky and abundant that they sometimes block the legibility of the letters, as for example the K and Y. Another identifying characteristic is the sharp contrast between the thick and thin strokes as is obvious in the L, M, V, W, etc.

Most of the letters occupy the area of a wide rectangle to give the effect of a squatty expanded styling. The I and J are obvious exceptions.

■ ESSENTIAL CONSTRUCTION

In laying out copy in this alphabet, allow generous space for the letter shapes. Not only is each letter extended basically, but additional space must be allowed beyond the base form for the wide serifs which extend well beyond the main verticals, horizontals and diagonals. Although the general effect of this alphabet is that of solidity, the variation in the thickness of the letter forms is a constructional requirement to make it possible to get the basic strokes within a given amount of space. This is especially obvious in the S, but applies to practically all other letters as well. Latin Wide may be rendered in comp or flat pencil technique, chisel-edged showcard brush or the more refined tools of reproduction lettering.

■ ESSENTIAL APPLICATIONS

Reduced to a small scale, this style is very appropriate for letterheads, labels —provided the wording is kept to one long line, acting as a contrasting foil to a frilly or flourishy script. On a large scale, this bold and brash alphabet is appropriate for display panels for exhibits, theatre lobby work, and well designed truck lettering layouts. Because of its constructional rigidity, you'll find Latin Wide easy to cut out of wood, beaver board and other three-dimensional material.

ABCDEFGHIJKLMNO
PQRSTUVWXYZ
1234567890&!?$
abcdefghijklmnopqrstuvwxyz

ESSENTIAL CHARACTERISTICS

The type face shown on the opposite page is easily identified among the hundreds of type faces in use today. At first glance it seems like a Persian alphabet with its intricate calligraphic twists and turns and variations in slant. Note that the O is slanted, while the U is almost completely vertical. This inconsistency is characteristic throughout the entire upper case alphabet. Likewise the capital letters vary somewhat in size. Compare the size and restraint of the F or the U with the wild flourishes of letters such as the B, D, L, Q, Y, etc.

The lower case letters are somewhat more consistent in size and slant. A definite characteristic here is the height of the ascenders and descenders and the extreme contrast of thin and thick elements of the letter strokes.

ESSENTIAL CONSTRUCTION

This is one of the most calligraphic type faces which have become standard in contemporary typography. The most appropriate tool required to render Legend is a broad-nibbed pen, carpenter's layout pencil and short-haired well-chiseled lettering brush. Whichever tool is used, it must be held rather rigidly, without rolling it between the fingers and freely using the side and full width of the instrument in a natural manner, as in normal handwriting rather than formal lettering.

ESSENTIAL APPLICATIONS

Because of its Persian appearance, Legend is a natural choice of alphabet for things exotic, oriental and mystic. Capital letters are not legible when used to form complete words. Used with discretion, Legend is an ideal alphabet for letterheads, labels, packaging for perfumes and cosmetics, chapter headings of story and book publication and the like. It does not rank high in legibility and therefore common judgment would dictate that this type face be reserved for display where emotion and sentiment are more important than urgency or objectivity.

abcdefghijklmn
opqrstuvwxyz
&!?$1234567890

ESSENTIAL CHARACTERISTICS

This type designed by one of Holland's leading type designers, at first appearance may seem like an erratic attempt at lettering. There is a seeming potpourri of lower case letters with upper case and the curves lack roundness. Libra must not be judged mechanically by the cold eye of a perfectionist who seeks obvious consistency, but by an aesthetic calligraphic-oriented person. Libra is a standard type face but what is so unique about it is that it does *not look* like a type face. It shows the strong influence of manuscript writing, from which it was derived. Libra is a "biform"—which means that there are no upper and lower case letters as distinct fonts. The A for example is lower case in construction, and so is the D, M, N, etc, yet the B, G and R are distinctly upper case. The body height of all letters is maintained throughout the alphabet, ascenders and descenders going above or below the imaginary guide lines.

ESSENTIAL CONSTRUCTION

This is a pen letter type which clearly shows the natural stroke direction of the lettering tool. A broad-nibbed pen held somewhat rigidly in the manner of a pencil should be used, or a well flattened-out chisel brush held in the same manner. To retain its true calligraphic character, all letters should be rendered with a minimum of strokes, preferably (where possible) without lifting the tool off the paper even for change in direction of strokes. The juncture of strokes should not be artificially rounded off. Let the strokes meet and stay as they naturally fall into place.

For purposes of the hand letterer this type face should merely serve as an aid. It should not be copied slavishly. Let the tool decide its natural direction—don't harness it to follow an unwilling track. A true calligrapher will yield to the tool in the various combinations of letters and may make two identical letters slightly different—even in the same word.

ESSENTIAL APPLICATIONS

Libra, as well as other calligraphically-derived letter forms, looks best when it is small. Although this alphabet comes up to 60 points, it is the smaller sizes which look best. Each letter in isolation appears abrupt and twisted, but these apparent irregularities diminish when the letters are combined to form words.

It is best to keep the letters close together in the formation of words. The finished word should not be viewed as a group of individual letters strung together, but rather as a combined word image.

Libra is a fine alphabet choice for diplomas, certificates, and other formal documentaries. It is a good alternate to the overworked Old English and made more readable too!

ABCDEFGHIJKLM
NOPQRSTUVWXYZ
&?!$1234567890
abcdefghijklmn
opqrstuvwxyz

ESSENTIAL CHARACTERISTICS

A very beautiful and ornate type face, Lilith is easily distinguishable because very few other faces in the typographer's catalogue remotely resemble it. It is a sort of hybrid between Roman and italic, since all letters tilt somewhat toward the right. It has retained the feeling of handwriting in the modified flourishes which are faintly reminiscent of Spencerian script. Lilith's most predominant characteristics are the architectural floral-shaped decorative motifs which break symmetrically through the center of most heavy strokes. Another outstanding personality of Lilith is the combination of a heavy and thin outline shell, the white areas of which are textured with short horizontal lines equally spaced, but which do not go through the floral pattern.

While the lower case letters obviously belong to the same family as the upper case, the ornamental floral pattern has been eliminated. Of special structural interest are the split terminals of the f, l and t, though this design motif is not maintained throughout the other letters.

ESSENTIAL CONSTRUCTION

Lilith is a standard type face and is available in printers' type, photo lettering and paste-up lettering sheets. However there are occasions when none of these are expedient and it is necessary to render it by hand, either for comps or for showcards.

In all ornate lettering styles, or those with shadows and other embellishments, it is best to temporarily forget about these excesses in laying out the work. In this case too, Lilith should be laid out as if it were a thick and thin solid letter, of course retaining the profile of the type which gives it character. When that is accomplished, then is the time to go back and add doodles and spangles or whatever is necessary. As a showcard letter, a fairly good version of Lilith can be rendered with the use of the Speedball pen. Once the general outlines are pencilled-in with a fair degree of accuracy, it is not impractical to render a word or two by hand, although at best, it will demand more bench time than solid letters.

ESSENTIAL APPLICATIONS

This elegant alphabet is a good choice for things having to do with the theatre, music, literature, museum and art exhibits. Lilith has its commercial applications for showcards and displays advertising jewelry, expensive furs, glassware and other luxury items.

The predominance of fine hairlines must be a practical consideration when planning a printing job. Although as a printer's type face, Lilith is available in sizes as small as 24 points, it is hazardous to print that small unless the printing conditions are ideal. This means the right ink mixture, proper make-ready and care in the selection of a good printing stock.

An interesting and colorful effect can be achieved when the inside white areas are filled-in with one or two transparent colors. The effect would indeed be dazzling—but the occasion must be appropriate.

ABCDEFGHI
JKLMNOPQR
STUVWXYZ

ESSENTIAL CHARACTERISTICS

This unique type face which comes only in caps, is very much reminiscent of the heraldic inscriptions on ancient crests and shields. It is a blend of calligraphic writing and the illuminated ornate lettering seen on ancient manuscripts on parchment and fine paper. One of the chief characteristics of Lombardic, as indeed of most early manuscript lettering is the closed areas formed by means of thin lines which bridge across the elements of the letter form. You'll see this characteristic from the very first letter of the alphabet right through most of the letters. Note the connecting line closes up the C, E, K, M, etc. Another strong point of identification is the absence of completely straight lines in the main stroke. Examine the upright stroke of the B for evidence of this. Also note the consistent pattern of the quality of roundness, even in such letters as the E, N, U, W, etc. Note too the fine elongation of the serifs which are used in a very decorative ornamental fashion.

ESSENTIAL CONSTRUCTION

Whereas the standard Lombardic type face is shown here with a decorative inline within the main heavy strokes, this can be dispensed with if you are rendering this style by hand. The effect will in no way be diminished, and it will be much faster to get the job done by treating the heavy strokes as solids. In the hands of a proficient lettering artist, this alphabet in a modified form can be rendered as a one-stroke letter, using the full chisel width of the brush for the thick strokes and the side of the chisel for the thin strokes. For a more careful rendering, two different size brushes or lettering pens may be used. For casual showcards, a broad-nibbed pen will lend itself to the calligraphic effect which characterizes this unique alphabet.

ESSENTIAL APPLICATIONS

If King Arthur of the Round Table had personal stationery designed for himself, the chances are that the artist would have favored this type design. However, Lombardic Initial shown here has sufficient current applications to deserve a place in the type specimen books of many of the best type houses. It is an excellent letter to use for books, theatrical advertisements which deal with things biblical, royal or archaic. It is also an appropriate letter style for diplomas, proclamations and testimonials. With the thin inlines and connecting hairlines removed or modified, this alphabet can be posterized sufficiently to give it enough structural solidity to be shaped in three-dimensional forms, made of wood, metal or beaverboard.

LOMBARDIC INITIAL

145

ABCDEFGHIJKLMN
OPQRSTUVWXYZ
1234567890$
abcdefghijklmnopqrstuvwxyz

ESSENTIAL CHARACTERISTICS

This type face of distinction and classic perfection was created by the famous type designer and poster artist, Lucian Bernhard. The unique features which serve to identify Lucian Bold are its tapered hairlines which take on weight as they blend into the thin wispy serifs, as shown in the capital A, M, N; the above-the-guide-line meeting of the V-shaped element of the M; the beautiful curlecue flourish of the tail of the Q; the unusual construction of the W. The lower case has its own points of identification: the ascenders are longer than the descenders (compare the b and p); the ball terminals of the a, c, f, g, etc.; the rather dominant upper portion of the g and the tapered diagonal hairlines of the v, x and y.

ESSENTIAL CONSTRUCTION

It takes a lot of experience to render this style in such a way as to retain its subtle charm. There is great sensitivity in the tapered serifs and the delicate balance between thick and thin, to which the artist must be sensitive. For reproduction purposes the letter must be drawn in very accurately, before filling in with paint or brush. A modified version using a freehand technique may be achieved which will be acceptable for showcards and window signs. Though of course such casual freehand rendering will not be a facsimile rendering of the type as it is shown here, the delicate variation of thick and thin can be retained, as well as the tapering of thin strokes—with continued practice. The thin wispy serifs are natural terminals for the pen, when it is stroked left to right with the thin edge. For the lower case, rule in the high ascender line and low descender line, to conform to the model alphabet shown here. With practice, you will develop speed and the inherent character of this aesthetic type face without the continued need for copying it.

ESSENTIAL APPLICATIONS

This type face because of its classic restraint and subtle charm is best suited for printing jobs that reflect a classic mood. It is available in sizes from 10 to 72 points. Lucian is a good type face to experiment with for book titles, chapter headings, title pages, book jackets and the like. Commercially it has been used with excellent effect for advertisements of such varied products and services as high-priced ladies' accessories, expensive wines and liqueurs, brochures and direct mail pieces for better-class department stores, banks and other institutions. Lucian will add distinction to letterheads, package designs, and business cards.

Lucian Bold

ABCDEFGHIJKLMNOP
QRSTUVWXYZ
&!?$1234567890
abcdefghijklmnopqrst
uvwxyz

■ ESSENTIAL CHARACTERISTICS

This popular sans serif type face, related to the Roman in its thick and thin strokes has a distinct calligraphic quality which makes it easily identifiable. As in all calligraphic art, the tool is used with a directness that adds a distinct virility to the final result. Note the two lobes of the B. Each lobe represents a complete stroke and there is no attempt to smooth the transition between the two. There is an abruptness to the strokes which shows that the lettering tool was given force rather than subtle direction. The abrupt breaks are noticeable in the C, D, G, J and especially the O. Note also that the stroke endings are angular, as in the G, E, F, etc. Other earmarks of Lydian are: the cross over the right side of the A; the slightly splayed M; the extension of the diagonal of the N; the construction of the W made by two overlapping V's.

■ ESSENTIAL CONSTRUCTION

The chisel-like quality of the strokes suggests the use of a broad-nibbed blunt instrument. To attain this characteristic, a wide pen or a well chiseled flat showcard brush is used. For pen lettering, the thickness of the broad strokes of the letter should be equivalent to the spread of the nib of the pen so that the heavy element can be done in one stroke. To get the light stroke, the pen is used slightly at an angle. The same technique applies to the use of the brush. Both the thick and thin elements are done with one stroke. For the thin strokes, the brush is also held somewhat slightly at an angle. The heavy strokes are accomplished with the full chisel of the brush and with added pressure. Lydian as a rule is not a built-up letter. Where facsimile reproduction is required, Lydian repro proofs can be ordered from photo type services.

■ ESSENTIAL APPLICATIONS

Though its primary use is in newspaper and magazine advertisements, this type face can with practical modifications be employed for work that is produced by hand. It is an easy letter to do and can be done quickly with brush or pen. Lydian is a good alphabet to add to the display man's repertoire of type styles. When used as a display face, its use is confined to the upper case. Lydian may be found on book jackets, letterheads, labels, package designs, etc. For limited body text, the lower case can be used with assurance of legibility. However, where large amounts of copy are required as in explanatory copy in books, Lydian is not recommended because the abruptness of its curves makes for staccato reading. Lydian is excellent for all printing processes and needs little special handling in production.

ABCDEFGHIJKLMNO
PQRSTUVWXYZ &!?$
1234567890
abcdefghijklmnopqrstuvwxyz

■ ESSENTIAL CHARACTERISTICS

There is an angularity to this cursive alphabet which distinguishes it very readily from other cursives or scripts. This characteristic is even more evident in the lower case letters than in the caps. Another outstanding characteristic is the calligraphic twists and turns which show evidence of a broad-nibbed or chiseled tool. Lydian Cursive was designed by one of the most renowned contemporary calligraphic artists, Warren Chappell, who also created the Lydian Italic and Lydian Normal shown on pages 153 and 149. Though the three Lydians all share the same calligraphic heritage, each one is distinctly different and has a personality all its own.

The Lydian Cursive caps do not line up mechanically because of the free flowing swashes which sweep below the guide line. Based on handwriting, this alphabet exhibits many variations in slant, yet visually all letters seem to be tilted in the same direction. The letters T and Y clearly indicate the license the designer has taken in the matter of the letter angle. The lower case shows greater mechanical regularity in this regard.

■ ESSENTIAL CONSTRUCTION

To duplicate this alphabet by hand, it must be rendered with the same calligraphic tool that Chappell used in designing it; a broad-nibbed pen held at an angle, making use of both the full width of the nib as well as the side of the pen. The width of the stroke is not achieved by pressure of the pen, but rather by using the broad side of the pen held at a 30° angle. The thin line strokes are produced by using the pen side or edgewise. All letters would be done with a minimum of strokes with no "doctoring" or artificial refinements. The thin strokes can be made by either going up with the pen, from bottom to top, or the reverse. Thus, in doing the A the thin strokes may be slanted at the bottom going up in one sweep, or at the apex. There is no set rule on the direction of the letter strokes in this or many other calligraphic styles, except the cross strokes which should be done from left to right.

■ ESSENTIAL APPLICATIONS

Most calligraphic lettering styles are traditionally relegated to "non-commercial" uses, such as for testimonials, diplomas, etc. Not so with this alphabet. Lydian Cursive has been in wide commercial use since 1940 when it was brought out by the American Type Foundry. Lydian Cursive is a versatile alphabet which lends itself amiably to book jackets, newspaper and magazine advertising, packaging and displays. While it is at its best a headline letter, it is safe to use it with moderation for limited body copy as well. Because it is a foundry type and therefore has to be hand set, discretion must be exercised in limiting its use for extensive body copy. Point of reminder: Do not use this (or any other flourish or swash calligraphic lettering style) in caps only. The caps should be reserved for initial letters only, to start off a line of copy with a flourish. Let the lower case do the real job in lettering your story.

ABCDEFGHIJKLMN
OPQRSTUVWXYZ
abcdefghijklm
nopqrstuvwxyz
¡!¿?$1234567890

ESSENTIAL CHARACTERISTICS

Although Lydian Italic is a standard printer's type face (and a companion to Lydian) it has all the earmarks of a hand-lettered alphabet. It has a casual informal appearance and makes no attempt to conceal the abrupt but natural joinings of the strokes produced with a broad-nibbed pen. Note the C which distinctly shows the joinings of the two curving strokes at the top left side. This is typical of all round strokes. Other identifying marks are the overhanging apex of the A and N, the diagonal terminals of all upright and cross strokes, the lower case U, the sloping sides of the M, etc. Although the upper case alphabet shares identifying characteristics with the parent Lydian, except for the slight italic slant, the lower case alphabets differ from each other. The lower case letters of Lydian Italic have more of a handwritten style, for example the a, f, u, v, y.

ESSENTIAL CONSTRUCTION

This is a sans serif thick-and-thin alphabet with two distinct stroke weights. Although classified as an italic, the slope is very slight; actually just tilted enough to conform to the diagonal nib of the lettering pen. Although essentially a pen letter, Lydian Italic lends itself admirably to brush and paint. A well-paletted shorthaired showcard brush will produce the same structural characteristics as the pen. Since it has no serifs or other embellishments, this lettering style is ideal for very fast showcard work. In the hands of an experienced letterer, each stroke is final and should require no touch up or "doctoring" of any kind.

ESSENTIAL APPLICATIONS

As a standard type face, Lydian Italic is frequently used on book jackets, newspaper and magazine advertisements, posters, etc. Because Lydian is a simple letter, and in its italic form a quickly rendered alphabet, it is very frequently employed for showcards, paper signs, oil-cloth banners and the like. This alphabet is one which you will want to use not only for a deluxe job, done rather carefully to suit the needs of the occasion, but it also has the advantage of speed for "knocking out" a "quicky" when time is of the essence.

ABCDEFGHIJKL
MNOPQRSTUVW
XYZ1234567890

■ ESSENTIAL CHARACTERISTICS

Here is a foreign import which has gained in popularity among American typographic layout artists. Designed in Italy in 1939, it has been widely adopted in America and is now available in a wide range of point sizes and in a number of variations in weight and letter width. Although it may appear at first glance as a square letter form, a closer inspection will reveal that actually letters like the C, O, R, S, etc., are not mechanically square. It is this subtle squareness which gives Microgramma its special continental character. The O, for example, is not a mechanically ruled-in rectangular area and the sides of this letter have a very subtle curvature, not only at the corners but the vertical strokes as well. This becomes more evident if you analyze the inside or counter of the letter. Foreign in appearance is the construction of the K which you will note has a little cross stroke which extends to meet the two diagonals. Another unique characteristic is the double thickness of the tops of the M where the V-shaped elements meet the two upright strokes.

■ ESSENTIAL CONSTRUCTION

Microgramma Bold gives the appearance at first sight as an alphabet which can be rendered almost entirely with ruling pen and straightedge, but as we have mentioned before, this type face is not quite that mechanical in structure. However, some of the letter forms can be produced without any freehand technique. The entire A, the upright and three cross strokes of the B, the upright of the D, all of the E and F, etc. lend themselves to the use of mechanical instruments.

Microgramma is a comparatively easy letter to remember, once you have made a good copy of it. It is one thickness throughout, all letters are nearly the same width (with the exception of the M and W) and the letters are easy to space. Spacing is made simple because of two structural characteristics. One, nearly all letters are of uniform width, then too, many letters are "closed in." This "closing in" is seen in the C, G, R, S, etc. and is a European format of a number of type faces designed abroad. Spacing is therefore made easy because each letter fits into its own box-like shape and there are a minimum of open letters.

■ ESSENTIAL APPLICATIONS

The rugged appearance of Micro gramma makes it an ideal choice for a message with masculine appeal. Its fundamental mechanical appearance conveys a feeling of structural stability of modern architecture, mechanical engineering, electronics and automation. It is in fact, a very popular type face for services and products which relate to automation and modern engineering. Microgramma Bold, shown here, or any of its other versions is also used extensively in modern packaging, exhibition and display work, and industrial design. Its rugged structural quality makes it a "natural" for three-dimensional work in wood, metal, masonite, etc. There are no vulnerable appendages to look out for in fabrication of the letters or in handling. For this same reason this type face will stand up very well in printing by silk screen or any other method.

You'll be keeping up with the times when you add this modern type face to your collection of practical alphabets.

MICROGRAMMA BOLD

ABCDEFGHIJKLM
NOPQRSTUVWXYZ

¡!¿?$1234567890

abcdefghijklmnopqrstuvwxyz

■ ESSENTIAL CHARACTERISTICS

This calligraphic alphabet, though it appears frilly at first glance will, upon a more careful inspection, reveal a rather strong virility. In the upper case C, note the rather unrefined upper lobe. There is no attempt to make that curvature smooth and "artfully" graceful. Instead there is a determined downward pressure with the flat side of the brush. This is equally true of the E, G, O and S as well as many of the lower case letters. This is a general trend in most of the present day "casual" or cartoon type of alphabet.

■ ESSENTIAL CONSTRUCTION

The handling of the brush will influence the effect you will get. As in most calligraphic alphabets, the brush grip here should be rather rigid and forceful. As to the particular type of brush, that will be a matter of personal preference. Though most traditional showcard writers are accustomed to work with a chisel edge rigger, illustrating artists who do lettering, use a pointed water color brush. In the lower case you will note that most of the letters are done with practically one stroke, without lifting the tool. This will become obvious if you follow the directional strokes of such letters as the b, f, h, k, m, etc. In other words, this alphabet is not lettered, it is written.

■ ESSENTIAL APPLICATIONS

This is not an easy letter to read on a passing vehicle, though it is legible enough when held at arm's length as in magazine or newspaper ads. It cannot be recommended for routine poster or billboard advertisement, and never use the caps alone to form words! Like the traditional Old English alphabets, the decorative nature of the capitals militates against legibility. Murray Hill Bold is a comparatively recent addition to the family of type faces. Whether it has enduring qualities—only time will tell.

ABCDEFGHIJKLM
NOPQRSTUVWXYZ
&!$1234567890

ESSENTIAL CHARACTERISTICS

Here's a type face you can spot immediately without difficulty. Neuland is the name of this alphabet and it is unique in many ways. Basically a Gothic style letter with the absence of serifs and delicate variations in stroke, it departs from this basic structure in the rough hewn effect produced by the angular shapes of both the "curves" and stems. The lobes of the B are typical of most of the other letters. Examine the D, G, P, R, etc. The feeling of letters cut in wood or stone is also evident in the straight strokes as well. These are slightly wedge-shaped and angular and reflect the intention of the designer and are not merely careless, as perhaps might appear to an untrained critic.

ESSENTIAL CONSTRUCTION

Although the sample here is the authentic type face used by printers, the hand letterer is at liberty, indeed he may be tempted, to experiment with individual variations, depending upon his tool, medium and job requirement. For pencil "comp" rendering, a flat chisel lay-out pencil will adequately serve the purpose to give the typographer the "effect" of Neuland without a slavish time consuming rendering. For hand painted work on display panels, showcards and other such uses, a short-hair flat brush held like a pencil (rather than twirled between the fingers) will easily give a fairly good facsimile of this type face.

ESSENTIAL APPLICATIONS

The pick-ax appearance of Neuland would seem to be off limits in merchandising dealing with the light touch. Neuland, of course, is very much in place when it has a man's job to do. What does that include? Smoking tobacco, tractors, men's wear, active hobbies and sports and that sort of thing. But Neuland is not limited to the rough outdoors. Strangely enough this type face goes well with art as well as industry and for this reason you'll see it utilized in advertising, art exhibits, book covers and concert posters.

ABCDEFGHIJKLMN
OPQRSTUVWXYZ
&!?$1234567890
abcdefghijklmnopqrst uvwxyz
abcdefghijklmnopqrst uvwxyz

News Gothic Bold

■ ESSENTIAL CHARACTERISTICS

This is one of the typographic "work horses" of the advertising field. "When in doubt, use News Gothic" is one of the commercial adages of the trade. Its recognizable feature is its simplicity. It is a sans serif letter form devoid of excesses and doodads. News Gothic as shown here is most suitable for display and sign work, since it is bold and conforms closest to the average man's mental image of the letters of the alphabet.

The characteristics of News Gothic Bold are: the appearance of uniformity of thickness (though the strokes are somewhat varied in thickness); the letter G which has a low cross stroke to which is appendaged a short vertical stroke; the point of contact of the lower diagonal stroke of the K as it meets the upper diagonal; the double weight thickness of the upper V-shaped element of the M (this also applies to the upper left element of the N); the unique Q in the elbow-shaped tail it carries. The lower case shows a more discernible variation in thickness of stroke, as can be seen from the m and n. The dots over the i and j are square-shaped.

■ ESSENTIAL CONSTRUCTION

In laying out this style in pencil, preliminary to the finished rendering, it is best to think of News Gothic Bold as a type face of uniform thickness, and leave the refinements and subtle variations in weight as a last step before the actual painting-in of the letter. As a general rule, all horizontal strokes are somewhat thinner. So too are the north and south segments of curves, such as in the G, O, Q, S and U.

News Gothic Bold has a dual job function. As a type face used in printing it is a long time favorite among typographic artists and art directors. As a basic Gothic, it is used very widely (with variations) by the sign painting field. In the hands of an expert sign painter, this type face (or a practical facsimile thereof) becomes a one-stroke letter done rapidly with a minimum of wasted strokes of the brush and mahlstick.

■ ESSENTIAL APPLICATIONS

In its dual role as a typographer's favorite advertising type face and as the model which serves as a basic alphabet for the sign field, the News Gothic shown here is used extensively in all advertising media—from the slick magazine to the side of truck panels. This versatility, however, does not apply to the lower case alphabet which is generally reserved for newspaper and magazine advertising where it takes a secondary position to the caps which are used for headlines.

Because this type face, and its variants, (News Gothic Standard, Light, Condensed, Extra Condensed) are so much in demand, the basic type comes in a wide range of point sizes as a printer's type, as well as in many of the newer forms of applied lettering sheets, hot press type, ready-made cut out letters in wood, cardboard, metal and plastic materials. These ready cut three-dimensional letter forms are available in various colors, in a wide range of sizes and thickness of stock.

ABCDEFGHIJKL
MNOPQRSTUVW
XYZ

&!?$1234567890

abcdefghijklmnopqrstuvwxyz

ESSENTIAL CHARACTERISTICS

Here is a powerful poster alphabet which should prove a welcome addition to your roster of practical type faces. It is simple, direct, easy to do and easy to read.

In the main, this is a sans serif "gas pipe" type of Gothic, tilted at an angle for emphasis and fast reading. There are a number of inconsistencies in its construction which deserve special note. Though this type face may be loosely classified as a thick-and-thin letter style, the thin strokes vary in width, as for example the cross strokes of the A, H, L and T. The V, W, X and Y are practically uniform in stroke thickness, whereas the Z is decidedly thick and thin. The lack of structural consistency shows itself also in the lobes or round parts of the letter form. Compare the circular lobes of the B and P with the flattened circular element of the left side of the C or the compressed sides of the O.

ESSENTIAL CONSTRUCTION

This is a practical style to render in pen, brush or mechanical instruments. The choice of tools will depend upon the nature of the job, the degree of perfection required and whether the lettering is to be rendered with ink or paint. To get uniformity in angle, an adjustable-head T-square can be used to draw in angle guide lines. Or the stock on which the lettering is done may be tilted, taped on the drawing board, thus fixing the angle, in which case a standard fixed-head T-square can be employed.

ESSENTIAL APPLICATIONS

The bold structure of this type face suggests its use on advertising dealing with heavy industry, electronics and products relating to masculine wear and interests. It is an ideal style for exhibitions, displays and hand-painted lettering as well as for cut-out materials in board, wood and metal. In addition, Old Gothic Bold Italic is a good style to use on truck lettering. Not only is it easy to read close up as well as at a distance, but it is also readable on moving vehicles, especially since the slant of the letter suggests the forward movement of the truck. Even when the truck is standing still, the italic gives the illusion of forward motion and speed.

ABCDEFGHIJKLMNOPQRSTU
VWXYZ
&.,?!$1234567890
abcdefghijklmnopqrstuvwxyz

ESSENTIAL CHARACTERISTICS

This is an alphabet which seems to exude a spirit of cold formality. It is severe in the extremes of its elements—hair-thin strokes, heavy uprights and needle-fine serifs. Onyx is a condensed type of alphabet, tall and erect and meticulously precise. To be more specific: The counters (inside areas) of most of the letters are uniformly narrow. Note that the counters of the B, D, O, P, Q and R are narrower than the basic upright strokes. Other tell-tale marks of Onyx are the wedge-shaped endings of the E, F, G, T and Z and the triangular serifs of the A, M, N, U, V, W, etc. Another point of identification: While the outside lobes (round parts) of the B are round, the inner sides are flat. This applies to all other lobes. The lower case reflects the same characteristics.

ESSENTIAL CONSTRUCTION

To render a good hand-painted version of Onyx would be a ticklish job for any reproduction letterer. While Onyx is essentially an instrument-made alphabet, the job is none the easier simply because of the precise nature of its construction. Any marked variation in the hairline thin strokes would be easily discernible, and no compass can come to your aid in shaping the round elements. Where letterpress type or phototype do not fit the bill, then, sharpen your pencil and follow with patience this model. It is an extreme challenge to precise and thorough lettering artistry, and the accomplishment should provide much satisfaction.

ESSENTIAL APPLICATIONS

The delicate hairlines of Onyx make it inadvisable to use this style for posters which have to be viewed hurriedly. That and also the fact, as we said before, that it is a difficult letter to render by hand. Otherwise, Onyx can be used for most advertising purposes—book jackets, stationery, ads in magazines and newspapers, etc. When Onyx is reduced, the hairlines become somewhat too fine for clear reproduction. This is especially so in reverse printing, where the white of the paper forms the lettering. Thus, your art should not require any sizeable reduction. Both the printer and engraver who make the plate will thank you for this caution.

ABCDEFGHIJKLMNOPQRSTUVWXYZ

1234567890

abcdefghijklmnopqrstuvwxyz

ESSENTIAL CHARACTERISTICS

Peignot comes in two versions, both of which are shown here. The top version does not exhibit any unusual characteristics which depart very much from dozens of other sans serif thick and thin alphabets. Special attention is called to the Q which has a cross stroke cutting through the bottom center of the letter. The letter K is also somewhat different in its treatment of the diagonals which barely touch the main upright. The V comes to a point at the bottom, whereas the A is flat at the top. The W too ends in sharp points.

The bottom version of Peignot is known as a "biform." This means that lower case and upper case letters are used interchangeably. To the eye trained for consistency, this seems like a badly scrambled combination of letters, but "biforms" such as Peignot have gained in popularity among typographers both here and in Europe. In fact, it is this lower version of Peignot which is used more widely than the upper version.

ESSENTIAL CONSTRUCTION

The word biform signifies that both upper and lower case letters are interspersed. Actually then, there is no distinct lower case to serve as a companion to a distinct upper case. The body of the letter is of uniform height, but certain letters rise and fall above and below the "body" guide lines. The A which follows the standard structure of the Roman letter form, is actually a letter reduced in height to be contained within the area of the "body" guide lines. This oddity repeats itself throughout the alphabet. In addition to this peculiarity, there are structural notations that deserve attention. Note the round apex of the top of the A, an oddity which is repeated in the M, N, as well as the bottom of the V and W.

For showcard purposes, a Peignot biform effect can be achieved with a one-stroke brush technique for most letters. Take the L for example. Starting with the top of the main stroke, the brush is pulled down and adroitly guided around the corner to produce the horizontal. This can be accomplished where speed is essential without lifting the brush. Other letters too can be done with an economy of stroke, the degree of finish depending upon the requirements of the job.

ESSENTIAL APPLICATIONS

Peignot is a comparatively new entry into American typography. Its use is as yet reserved for sophisticated merchandising campaigns with the media of newspaper and magazine advertisement. Some national advertisers however have cautiously introduced Peignot (in its biform version) for truck posters, banners and lettering for industrial exhibits. It's still a bit of a shock to many people to read a line lettered in Peignot biform. But it is this element of shock which is precisely the reason why it is being used with increasing success.

The modified version of Peignot shown on the top line is a strong legible display letter and has, of course, wide acceptance for all types of advertising and merchandising media. Its solid form without structural impediments makes it a good choice for posters and displays—either in paint or a three-dimensional cutout. Its application also extends to all typographic needs of book publication, letterheads, business cards and the like.

Peignot, in both versions reproduces well. The elements are heavy enough to "hold their own" even under adverse conditions due to nature of stock or ink.

ABCDEFGHIJKLMN
OPQRSTUVWXYZ
&!?$1234567890
abcdefghijklmnopqr
stuvwxyz

ESSENTIAL CHARACTERISTICS

This is a Roman type at its best. The letter forms show both restraint and elegance. Each letter seems chiseled out of fine marble. Specific letters which identify Perpetua are the A with cut-off apex, the low-slung upright of the G, the fishhook appendage of the J which falls below the bottom guide line, and the slight diagonal sides of the M. Also note the rather wide structure of the Z. The serifs are thin but bracketed to the main stem and are an integral part of the letter.

ESSENTIAL CONSTRUCTION

It takes a skillful letterer to do a classic Roman alphabet, but in addition to dexterity and skill, the lettering artist must "feel" the structural beauty and balance of each letter as he shapes it. There is no hard and fast rule which will serve as a guide in constructing a type such as Perpetua. Though there is an affinity of each letter for the other, each letter of the alphabet has its own proportion and accent. There is considerable variation in width between letters. For instance, the C is wide while the L is narrow. The N is much wider than the S. What makes the letters of the alphabet hold together is the tone or "color" which results from the consistency in relationship between the thick and thin elements which are characteristic of this alphabet.

ESSENTIAL APPLICATIONS

The stately appearance of this type face suggests conservatism and classic beauty. It is an ideal type to use for a bank, library, a concert hall or public institution. This is a type which whispers, not shouts, and therefore it would be unwise to use it for blatant claims for a fire sale or bargain basement special. Perpetua is at its best when it reflects a product of merit, aesthetic values and permanence. It is a letter which carries the mark of sterling and purity of form.

ABCDEFGHIJKLM
NOPQRSTUVWXYZ
1234567890$

■ ESSENTIAL CHARACTERISTICS

Once you get to know the outstanding characteristics of Playbill you'll always recognize this unique type face. You'll remember it as "that condensed elongated letter with the thick, squarish-looking serifs." These squarish serifs are heavier than the main upright and diagonal strokes but of the same thickness as the horizontal strokes. All lobes and round elements are flattened on the sides but remain round on top and bottom as best exemplified in the O. Each letter is condensed so that the inside areas are narrower than the thickness of the upright basic strokes. It's look-alike among type faces is Barnum. Both are revivals of nineteenth century Egyptian condensed display faces.

■ ESSENTIAL CONSTRUCTION

Playbill is a mechanical type of letter, rather easy to put together, either with a brush or with ruling pen and straightedge. The letters are in the main uniform in construction. That is, the upright strokes are made the same thickness, and most letters occupy the same amount of space. To render this alphabet, first draw in the letters in pencil—all strokes of the thickness of the upright strokes. In other words, lay out this type as a condensed elongated one-thickness alphabet. When the preliminary pencil work is completed, draw two parallel lines, one on top and one on the bottom representing the thickness of the serifs. Make these guide lines go through all the lettering. Next, mark off the serifs and adjust the serifs and round elements to the thickness of the guide lines. Next, ink or paint-in, using a brush or ruling pen, depending upon the degree of perfection required. That's all there is to it.

■ ESSENTIAL APPLICATIONS

The name of the type face Playbill is symbolic of a nostalgia associated with the theatre "in the good old days." This is a modern version of the type face once so popular in France, where it decorated the theatrical placards on kiosks and boards. It was also popular in our own country at the turn of the century. Today, Playbill is used as a novelty type face on displays, printed advertisements, TV and movie titles. It adds an ultra modern and sophisticated touch to letterheads, labels and brochures, especially when it is used in combination with a script of a more conventional face. Playbill is an interesting type face to experiment with and should be used as a "one-line" face; otherwise it proves difficult to read. It is not available in lower case, but one could be improvised easily enough, based on the same structural format.

ABCDEFGHIJKLM
NOPQRSTUVWXYZ
&?!¡¿1234567890

■ ESSENTIAL CHARACTERISTICS

This is one of the easiest alphabets to identify. There is none quite like it in the entire repertoire of standard type faces. Essentially Prisma is a block letter composed of fine lines running parallel to each other and shaped to fit the contour of the basic letter form. In addition to this unique quality, there are other characteristics. The round letters are circle round, either in complete form such as the O, Q or partially so as in the C, S, U, etc. The sides of the M are diagonal and the S is a combination of two round elements exactly identical. The lower element of the J is flat and the W is an inverted replica of the M. But to describe these minor characteristics of this most unique of type faces is like pointing out the flower in the lapel of a suit worn by a two-headed man.

■ ESSENTIAL CONSTRUCTION

This type face is a draftsman's dream! It's an exercise in the use of compass and ruling pen. The mechanical structure of the letter form calls for accurate precision and control of tools. In a sense, each stroke of each letter is repeated five times, a rather time consuming business considering today's emphasis on speed. It is gratifying to know, however, that this type design (as well as most other standard type faces) is available in ready-made paste-on or pressure stick lettering, as well as in photo lettering in a generous assortment of sizes.

■ ESSENTIAL APPLICATIONS

Since Prisma has the quality of precision, it is an ideal letter to use for matters dealing with engineering, machinery and draftsmanship. However, it is not confined to such obvious uses. It has its place in general advertising for magazines, newspapers, brochures, annual reports and packaging. If it is limited to just a few words, Prisma can be a very effective type face and will serve as a good addition to your stock of "special" alphabets.

ABCDEFGHI
JKLMNOPQR
STUVWXYZ
1234567890

ESSENTIAL CHARACTERISTICS

This is a very popular revival of the old Egyptian slab-serif alphabets, very much in evidence as headings in modern newspapers and magazine advertisements. It has many decided characteristics which identify it very easily. In addition to the square-looking slab-serifs, Profil's strongest identifying feature is its multiplicity of outline, inline and heavy drop shadow effects. Stripped of these embellishments, Profil is a solid looking Roman thick-and-thin black-faced alphabet, surrounded by a white inline which in turn is held together by a thin outline. The shadow is deep, falls to the right and below, a quality which gives it a three-dimensional effect. Profil may be classified as an italic because of its forward slant, but it has no counterpart in a straight up and down Roman. In fact, this is one of the few current type faces which has no variation from the basic version shown here.

ESSENTIAL CONSTRUCTION

Without its embellishments, Profil is an italicized Roman, thick and thin with square serifs. The best way to construct this alphabet is to start off with the basic structure, and then build it up from there. You will note that despite its rather broad structural elements, the letters are condensed as if to limit the moderate amount of space it would wish to occupy. Note the flattened lobes of the C, D, G, O and Q. The round elements seem compressed in most cases, but this is not consistent with all letters, as for instance, the B, P and R which have the traditional round lobes generally associated with the Roman letter form.

ESSENTIAL APPLICATIONS

Profil is most decidedly a display rather than a text letter. In fact, it was not designed with a corresponding lower case. It is an excellent type face to use for brief headings and lends impact and color to any ad—especially if the body text is composed of a conventional type face such as Times Roman or Caslon.

Profil can be used with good effect on focal copy of letterheads, labels, newspaper and magazine ads, and direct mail. However, its use need not be confined to such media. It is a good alphabet style to use for truck and window letters, exhibition panels and headings for display boards for both interior and exterior display purposes.

ABCDEFGHIJKLMN
OPQRSTUVWXYZ
abcdefghijklmnopqrstuvwxyz
&!?$1234567890

■ ESSENTIAL CHARACTERISTICS

This alphabet designed by Tommy Thompson represents yet another variety to the many calligraphic alphabets which are improvisations of our best lettering craftsmen. The style shown on the opposite page is not intended as a master pattern for copying. It is offered here rather as an approach and source of inspiration for personalized experimentation. In essence, the mark of the lettering pen is in evidence in every stroke and swirl. The broad-nibbed pen, held at a rigid and unvarying angle will naturally produce the twists and turns which artifice alone can never duplicate.

■ ESSENTIAL CONSTRUCTION

The pen or well-paleted chisel brush will be called for here. The angle or slant may be varied to fit the needs of the letters in combination. Notice for instance, the somewhat upright Y compared to the more italicized Z. It is not that the slant of the italics must remain constant for all letters. A good calligrapher does not look at each letter individually. His vision encompasses the entire word or perhaps the entire page or panel of copy. It is the tone and relationship between letter and letter, and word to word that produces the total effect. It is difficult therefore, for us to analyze each letter form as if it were fixed in frozen form and should be copied stroke for stroke. Uniformity of effect rather than inflexible uniformity in structure is important here.

■ ESSENTIAL APPLICATIONS

Diplomas, certificates, testimonials, book jacket headings, invitations are some of the most obvious suggestions for using a calligraphic form. It must be chosen in good taste for an audience with sentiment and good taste. For appropriate use of this type face, the product advertised must be considered. Ads for cosmetics, perfumes and most luxury items would make excellent use of this type face. Quillscript is still another alphabet to add to your collection of calligraphic faces.

ABCDEFGH99IJKL'MMO
PQRSTUVWXYZ
&!?$1234567890

abcdefghijklmnopqrstuvwxy18

■ ESSENTIAL CHARACTERISTICS

Here is another connecting script, a bit more reserved than the other scripts shown on pages 22 and 62. The lower case alphabet bears a resemblance to Kaufmann Script which appears in this book on page 135. The lower case letters which are primarily used in the formation of words, connect to form words in the manner of handwriting—and are largely based on natural writing. These letters are somewhat more formal in their structure, thickness and angle than the caps. The thickness of strokes vary in width much more in the upper case than the lower case, as is evident by comparing the letters D and d for example. There is considerable structural differences among the upper case letters to establish a consistent pattern to aid in identification. The A and G for example, are simple and give more the appearance of lettering than handwriting, whereas the D, F and J are distinctly calligraphic.

■ ESSENTIAL CONSTRUCTION

Like all alphabets based on handwriting, Repro Script is best rendered if the tools and technique of natural handwriting are employed. The tool may vary from the standard writing pen. It may be a broad-nibbed pen of sufficient flexibility to allow for variations in stroke pressure. A felt tipped brush-pen may be used and will allow for greater speed since it is self-inking. Where opaque color is required for coverage on a dark background, a standard showcard brush will have to be used. The bristles would be short, come to a ready chisel and "springy" enough to flex back after pressure is released.

Again, like in most calligraphic lettering, the artist should not attempt to slavishly follow the model alphabet like a mechanical pantograph. The model alphabet should merely serve as a guide showing direction and effect, and nothing more.

■ ESSENTIAL APPLICATIONS

An informal script should not be burdened with a lot of copy, a caution that applies also to all italics and decorative alphabets in general. The use of Repro Script should be held in check and be used only as an occasional headline. As such, it is fine on showcards, posters, direct mail brochures and newspaper and magazine advertising. Used with moderation, it expresses urgency and commands attention as an effective eye catcher. It lends itself to other than straight horizontal layouts—it can be very effective if placed on a slant, running uphill, often punctuated with an exclamation mark. Example: For a limited time only!

Repro Script (or your version of it) lends itself very adequately to quick lettering with brush or pen. You'll be glad to make use of it at the right occasion.

Repro Script

ABCDEFGHI
JKLMNOPQR
STUVWXYZ
&1234567890

■ ESSENTIAL CHARACTERISTICS

This highly ornamental type face is really a glorification of the Roman dressed up with baroque flourishes reminiscent of the golden age of the past. A closer and less sentimental analysis of this style will reveal the consistency of its structure and ornamentation. The serifs of the A, you will note, are symmetrical split curves going in opposite directions. This is also true of the bottom serifs of the F, H, I, K and several other letters. In fact, the lower half of each letter is considerably different than the upper part. Whereas the upper structure is a rather conventional block-serif letter, the lower half has most of the embellishment. The decorative element which gives unity to both upper and lower half is the graduated parallel lines varied in spacing to give the illusion of blended shading.

■ ESSENTIAL CONSTRUCTION

This exotic type face is not common, and only printers who make a specialty of exotic or "old-fashioned" antique faces carry it in stock. If a hand rendered facsimile type is needed, it can be achieved by laying out the basic thick and thin structure first and then adding serifs, drop shadows, curlicue embellishments and last of all, the graduated line effect within the letter. It would be a mistake to incorporate all these embellishments at one attempt. Built up from its basic Roman structure and *then* adding the ornamentation will greatly simplify not only each letter as you make it, but you will be able to do the job faster and with greater feeling for spacing.

■ ESSENTIAL APPLICATIONS

The ornamental nature of this letter form suggests its main uses: Posters and advertisements dealing with subject matter of antiques, art exhibits, literature, expensive silverware and furs, book jackets, titles for romantic stories, and other subjects reflecting elegance, royalty, luxury and intrigue. There are several variants of the Romantique type face. The model shown on the opposite page is called Romantique No. 5. Add this exotic lettering style to your collection of special alphabets for special occasions.

ABCDEFGHIJKL
MNOPQRSTUVW
XYZ
&!?$1234567890
abcdefghijklmnopqrstuvwxyz

ESSENTIAL CHARACTERISTICS

Rondo Bold is a type design based on free-flowing calligraphic writing. Like all calligraphic styles, Rondo shows unmistakable marks of the tool which is used in its construction—be it a blunt flat chisel-edged instrument such as the cutter's chisel, the carpenter's pencil, the flat-nibbed pen or the short-haired chisel-edged lettering brush.

The upper case letters of Rondo Bold are handled somewhat differently from those of the lower case. The capitals show more vigor, greater freedom and are somewhat less consistent than the lower case letters. Notice that most upright strokes of the caps are thin; whereas the lower case letters have heavy upright strokes. This heavy stroke is in general consistent throughout the alphabet, with the possible exception of the f.

ESSENTIAL CONSTRUCTION

When rendered by hand, it would be a mistake to slavishly copy the Rondo example shown here. To do so would rob this alphabet of its spontaneous charm. It is best to practice the manipulation of the tool (pen, pencil or brush) before proceeding to create the finished letter form. Do not rotate or twist the lettering tool as you would normally in constructing a Roman alphabet. Move your entire wrist and hand but keep the lettering tool gripped in a fairly rigid position. Study the style carefully, analyze it before you begin the actual lettering; but then do your lettering rapidly even if the results are not always predictable and uniform. Keep the rhythm and movement, work in harmony with the tools—the result will take care of itself.

ESSENTIAL APPLICATIONS

For sustained copy matter, use the lower case alphabet. The capitals used by themselves to form words make for poor readability. Rondo is a fine style for diplomas, testimonials, greeting cards and for other occasional purposes. It has the advantage over the traditional "Old English" style in so far as the type is easier to do and much less time consuming to render. What is more, it is generally more legible to the average person. Rondo, however, is not too suitable for general poster work as it lacks the punch of such letter styles as Futura, Neuland, Cooper, etc. This style is a good one to use for showcards and is one of the easiest alphabets to cut in stencils because of the freedom it permits in variations.

ABCDEFGHIJKLM
NOPQRSTUVWXYZ
&!?$1234567890
abcdefghijklmnopqrstuvwxyz

ESSENTIAL CHARACTERISTICS

Here is a choice alphabet for bold poster work. It is powerful, comparatively fast and simple to render, and easy to read. Classified as a sans serif type face, Samson has its own personality which makes it different from most other sans serif styles. Let's analyze some of these outstanding characteristics. Note the overhanging heavy stem of the A, which protrudes beyond the natural apex. You'll also find this point of identification in the B, M, N, P and R. Also observe the slight diagonal top and bottom ending of most upright strokes. The letters H and I are typical, but you'll find it with consistency in most heavy strokes. Note the intentional abruptness of the curves, especially obvious in the counters of letters, as in the C, D, G, O, etc. These give this alphabet a distinctly calligraphic effect. The lower case and numbers share these characteristics.

ESSENTIAL CONSTRUCTION

The calligraphic effect is accomplished by using a minimum of strokes, with broad lettering tool, which may be a diagonally-nibbed lettering pen, a short-haired well-chiseled lettering brush or a layout pencil cut with a diagonal chisel. Each stroke is distinct, definite and final. Most any letter here can be done in two or three strokes, with no need to "doctor" the stroke or introduce refinements. Indeed, it is the decisive quality of this style that gives it power and impact. To get the abrupt twist in the round strokes, the lettering tool should be held rigid and fixed between the fingers, and *not* twirled as is customary in the traditional Roman lettering style. Even the dot over the lower case i is a short stroke, without any semblance of curve.

ESSENTIAL APPLICATIONS

The most obvious application of this style is for poster work, especially those posters dealing with the theatre, motion pictures and other theatrical events. This letter style also has some of the flavor of the romantic and heroic past reminiscent of the old and golden days of knighthood, chivalry and swash buckling gallantry. It brings to mind the exploits of the Three Musketeers, Robin Hood, and piracy on the high seas. More contemporary applications of Samson are for ads dealing with art exhibits, men's wear, service industries, men's cosmetics, etc. And it's a cinch to cut in three-dimensional material such as wood, beaver board, as well as for hand-cut film stencils.

ABCDEFGHIJKLM
NOPQRSTUVWXYZ
1234567890

ESSENTIAL CHARACTERISTICS

Here is a type face which sparklingly reflects its well-chosen title, Sapphire. A Roman alphabet, with refined well-modulated serifs, its most identifying clue is the diamond-shaped floral pattern which serves as an inset within the heavy strokes of each letter. Additional clues of identification are the compressed lobes of the C, D, G, O and Q. This is not a consistent characteristic since the lobes of the B and R are fully rounded, as are the compound curves of the K and the clipped tail of the Q. Note, too, the contrast in curvature or semblance of a waist line in practically all strokes, especially in the thin lines such as the left diagonal thin stroke of the K and thin elements of the M, N, etc.

ESSENTIAL CONSTRUCTION

When rendering this style, either in the initial layouts or the finished letters, it is best to momentarily forget the embellishments within the heavy strokes. Of primary importance, both in achieving the right weight and feeling of good spacing, is the structure of the basic letters, with special emphasis on the relationship between thin and thick elements. Get the serifs just right, place each letter squarely and proportionally within the guide lines, concentrate on the basic profile of the letter. When all that is done it is a simple matter to add the embellishments, almost as an "afterthought." The embellishments should never structurally interfere with the basic letter form. In other words, finish the entire lettering job first and then, and only then, introduce the sparkle elements.

ESSENTIAL APPLICATIONS

Where is the Sapphire type face, shown here, most appropriate? Services and products and events which have to do with things feminine, dainty or costly. You may want to refer to this alphabet next time you are doing a poster for a flower show, or an ad promotion for furs, expensive dishes, jewelry, ladies' wear, good wine, sparkling champagne or for a fine arts exhibit or beauty salon display. As an added bit of colorful lustre, you can enhance the effect further by introducing one or more colors to the embellishments. You will get a luminosity which will earn you many compliments. One bit of advice: For best effects of this exotic type face, limit its use to one or two words or short headlines. Use another and more simple lettering style for massive lettering copy. It is only by discrete contrast of style, that you will achieve a well-balanced effect in any advertisement. Use highly decorative type faces such as Sapphire in moderation.

ABCDEFGHIJKLMN
OPQRSTUVWXYZ

&!?$1234567890

abcdefghijklmnopqrstuvwxyz

Spartan Heavy

■ ESSENTIAL CHARACTERISTICS

Here's a virile alphabet that asserts itself in any crowd. Essentially it is a Gothic one-thickness letter without frills and subtleties. Note the even length of the cross bars of the E and F, the perfect circle C, G, O and Q, the diagonal side strokes of the M and the rather wide W. The clues to the lower case identification are the perfect circle "inside" areas of the a, b, d, g, o, p and q, while the "outside" slopes of these same letters (with the exception of the o) are not circular concentric with the inside areas. The lower case alphabet departs from the uniform thickness of the upper case alphabet. For example, note the tapered curves which shape the a, c, d, g, m, n, etc.

■ ESSENTIAL CONSTRUCTION

For reproduction, Spartan Heavy is a mechanically constructed letter allowing full use of measured-off basic strokes, ruling pen, compasses and fill-in brush. It's an easy alphabet to construct by memory, because of its uniformity practically throughout. In the upper case all letters can be made the same thickness; round letters are rendered with compass. There are no subtle shaded variations in curves, no serifs. A more basic letter can hardly be found among the letterer's repertoire of type faces.

■ ESSENTIAL APPLICATIONS

This is one of the "standby's" of reproduction letterers, showcard writers, exhibit designers and typographers. Spartan Heavy is a "natural" for poster work, for stationary or moving displays. It is excellent for 24-sheet posters and exhibit panel displays. It looks good painted, cut out of wood, panel board, etc. It's good to know too that this very popular alphabet is available not only in type, but in photo lettering as well. It is also available in ready-to-use cut-out three-dimensional letters of wood, plastic, metal and various synthetic compositions, in many sizes, painted, unpainted, porcelain enameled and in a wide assortment of thickness.

ABCDEFGHIJK
LMNOPQRSTU
VWXYZ
1234567890
&?!$.,:;

ESSENTIAL CHARACTERISTICS

Here is a type face which doesn't have to wear its identification badge to be recognized in any crowd of type faces. Its name, Stencil, is accurately descriptive of its pedigree. This style available as a letter press type face and in photo lettering versions, is distinctively reminiscent of the early shipping case type of stencil used for identification, street signs, and simple wall stenciling. Stencil's strongest characteristic is the bridge gap or "ties" which break up the letter into integral parts. Essentially it is a thick-and-thin letter with short, stubby and rounded serifs. As a type face, it comes in upper case only, but a lower case can be improvised very easily when required.

ESSENTIAL CONSTRUCTION

Where for one reason or another expediency dictates that this type be lettered by hand rather than to make use of its standard form in type, it is best to construct this style without first thinking of the gaps and bridges. "Make the letter complete and connected," advise some experts, and then when that part of the job is finished, introduce the gaps by painting over the junctures where needed. In most cases this advice is good, because you will get better continuity in basic construction if at first you leave out the gaps. Unless the letters are drawn out in pencil very carefully, the introduction of these gaps may actually impede your progress, both as it applies to time and to final result.

ESSENTIAL APPLICATIONS

Stencil is a popular letter for occasions requiring a masculine and rustic appeal. Paradoxically though, we associate this style with inscriptions such as "Fragile," "Handle with Care," etc, but that, no doubt, is not so much due to the delicate sentiments as to the prosaic fact that such markings appear on shipping crates and freight baggage. Stencil is also reminiscent of the days of yore, of the type of announcements found in the Old West denoting a search for some colorful desperado whose apprehension was to be rewarded by a tempting prize intended to induce anyone to join the searching posse. In spite of these colorful associations of days gone by, Stencil type would not be included in this series unless it had many applications to present uses in brochures, newspaper and magazine ads, and other advertising jobs.

abcdefghijklmnopqrstuvwxyz

1234567890

ESSENTIAL CHARACTERISTICS

Stradivarius echoes symbolically the soft violin strains of hearts and flowers music. Indeed it derives its name from the famous violin maker whose craftsmanship has never been surpassed. Stradivarius bears a slight resemblance to the round hand version of Spencerian script—a form of writing that was the pride of most educated people of an era gone by.

The outstanding characteristic of Stradivarius is the flowing rhythm of hair fine swirls, accented heavily in the parts of the stroke which received the pressure of hand and pen. These pronounced heavy strokes act as a sort of counter weight to keep the swirls in balance. Note too the occasional ball terminal, disconnected from the body of the letter. This is noticeable in the H, K, N, etc.

The lower case letters are not like the caps, but they go well together when the caps are used as initial letters.

ESSENTIAL CONSTRUCTION

Though I have personally seen a showcard writer do a good replica of Stradivarius, directly with a brush, without careful outlines, it takes an inordinate amount of skill to attempt such a feat. He was using a pointed water color brush, handling this tool with as much ease as if it were a lettering pen. He had mastered this alphabet so well that he was able to use it on window display cards for a department store, where he was employed. And he was fast, too!

The recommended tool for rendering Stradivarius is a pen. The accented parts are made, not with a single stroke pressure of the pen but by painstakingly outlining each letter before filling in.

Stradivarius as a type face is available in point sizes 18 to 84 and is available in the better type houses.

ESSENTIAL APPLICATIONS

As a hand-lettered alphabet, Stradivarius would be appropriate for headline copy on fine window cards for department stores, haberdashers, jewelry shops, etc. The upper case letters cannot be used to form words and serve best when they are used as initial letters for words done in lower case.

The lower case letters are by comparison, rather simple. The long ascenders, like poised rifles on a row of soldiers, help to give the words rhythm, especially since the angle remains consistent throughout the alphabet.

If Stradivarius in any of its mediums (type, photo lettering, applied transfer sheets, etc.) is intended for reproduction, it will require the utmost care in printing, especially in small size reverse printing, as the thin hairlines would have a tendency to close up or disappear. Only the best stock is good enough for this beautiful delicate alphabet.

ABCDEFGHIJKLM
NOPQRSTUVWXYZ
&.!?$1234567890
abcdefghijklmn
opqrstuvwxyz

ESSENTIAL CHARACTERISTICS

This lettering style looks like oriental calligraphy. Note the three bold stab-like strokes of the A which cross over the intersections. This is a typical characteristic of most of the letters. The basic letter form of Studio is thick and thin, but the relationship of thick and thin does not follow a consistent pattern. The A for instance is distinctly thick and thin; however, the H and N are of one thickness. However, it is difficult, perhaps unnecessary to analyze each stroke of every letter to ascertain the characteristics of Studio. More important perhaps is the all-over effect achieved by this casual alphabet when the letters are put together to form words or word groups.

ESSENTIAL CONSTRUCTION

The charm of this alphabet lies in its unstudied casual freedom. The calligraphic effect is the result of the deft use of a well-chiseled instrument such as a carpenter's pencil, a square-nibbed pen, a short-haired chisel-edged brush or one of the new "Flow Master" fountain brushes. The instrument is held rigidly and is not rotated between the fingers. The letters should not be carefully drawn out in pencil outline. Instead the letterer would do well to use the model purely as an inspirational practice guide and make no attempt to copy the exact results. After a certain dexterity with the tool has been developed, the specimen shown here should not be referred to, except as an interesting basis for comparing the feeling rather than the minutae of construction of each stroke. If you evolve a variation so much the better. Try this style—it's fresh and practical too.

ESSENTIAL APPLICATIONS

Studio is essentially an alphabet having limited use. It may be substituted for Old English and Text to give diplomas, manuscripts, etc. a fresh modern look. Showcards for women's apparel are enhanced by the delicate quality of the lettering. Similarly, the dramatic feeling of the alphabet makes it appropriate for theatrical displays. Studio should not be used for continued copy and is best when viewed in a few words as a heading. It is a good style to use to break up the monotony of conventional type faces.

ABCDEFGHIJKLM
NOPQRSTUVWXYZ
abcdefghijklmnopqrstuvwxyz
&!?1234567890

ESSENTIAL CHARACTERISTICS

This is a standard type face not easily mistaken for any other and has a definite personality all its own. It is a cross between an italic and a script, but it derives its essential character from handwriting. This is evidenced by the interlacing loops of the A, B, P, etc. The heavy accented strokes, too, reveal its calligraphic character. Note the extension of some letters below the natural base line, as in the G, J, L, Q and Y—all reflecting the natural flourish of writing. Stylescript is a bold contrasty thick-and-thin alphabet, posterized to make it serviceable for commercial work on posters and displays.

ESSENTIAL CONSTRUCTION

There are essentially two thicknesses to this letter style, with modulating transitions between the basic thick and thin stroke. Though it appears as a free flowing lettering style, Stylescript is devoid of eccentricities or inconsistencies. All letters retain the same angle and relationship between thick and thin. Once the structure of this letter is understood and mastered, it can be rendered very fast, either with brush or pen. The variations and relationship between thick and thin will become as natural to the letterer as his handwriting and will after a while slowly modify itself to reflect the letterer's own personality.

ESSENTIAL APPLICATIONS

Stylescript can be used as initial letters for words done in upper case letters in combination with simpler alphabets. It can also be used to form complete letters, provided copy is restricted to single headlines, or single words. This is not a formal script, such as Thompson Quillscript and therefore need not be limited to occasions of elegance. Stylescript is a heavy duty work script lettering style which is applicable to routine copy on showcards for all occasions. It demands less attention than a more formal script and takes less time to do!

ABCDEFGHIJKLM
NOPQRSTUVWXYZ
&!?$1234567890
abcdefghijklmnopqrtsuvwxyz

ESSENTIAL CHARACTERISTICS

This version of Stymie gives the immediate impact of boldness and power. The italic slant of the letter adds a touch of swiftness to this forceful impact. Essentially Stymie Black Italic is a block letter with clubby square serifs, resembling the Beton family of type faces. On closer inspection, what at first appears like a one-thickness type face, shapes up into a letter with slight variations in thickness. For instance, the left side of the A is not as thick as the one on the right. This thick-and-thin characteristic follows through in the diminished thickness of all cross bars, curvatures and all diagonals. It is in fact even more discernible in the lower case. Identifying letters which are easily distinguishable from all other alphabets are the A with the top cross serif, the fore-shortened cross bars of the E and F, the apex formed by the two diagonals of the K.

ESSENTIAL CONSTRUCTION

This is a display face which lends itself to ruling pen or brush. It is a mechanical type of alphabet excellent for reproduction techniques. Though this alphabet, because of its mechanical precision, calls for accuracy in construction, the very boldness of its construction does not tax the letterer's fine sensibility and craftsmanship as does an alphabet such as Trafton, Weiss or Bernhard Roman. Because of the popularity of Stymie, this type face is available at most typographic composing rooms as well as in ready-made letters in transparent or opaque tabs intended for paste-ups. In addition, this type face can also be had in photo lettering.

ESSENTIAL APPLICATIONS

This is an excellent type face to use on industrial displays and exhibit panels. Its boldness commands authority and respect. Because of its rugged construction and complete absence of hairlines and vulnerable points of breakage, Stymie Black is an excellent choice for cut-out materials such as wood, foam ice, wall board, etc. Discretion would dictate that the most appropriate use of this bold type face would be in the fields of industry, men's wear, heavy machinery, electronics, etc., rather than the flimsy feminine precincts of lingerie, cosmetics or other such fluff fluff. It is good to know that this type face is as popular with sign painters for window displays, exhibitions, theatre panels, and other areas where hand lettering is used, as it is with layout artists and typographers for letterheads, packaging, book covers, etc.

Stymie Black Italic

ABCDEFGHIJKLMN
OPQRSTUVWXYZ
&!?$1234567890
abcdefghijklm
nopqrstuvwxyz

ESSENTIAL CHARACTERISTICS

Think of a one-thickness mechanically precise letter, add serifs of the same thickness, and you have a good concept of what Stymie Medium looks like. Now let's take a closer look at this very popular type face. The C, G, O and Q are mechanically perfect circles or parts of a circle. The cross serif on the top of the A is another identifying element, as is also the below-center cross bar of the G and the little appendage on the lower right side of this letter. The R is somewhat unique in the fact that the traditionally diagonal stroke is here straight and vertical. The cross bars of the E and F are short and are finished off with the same square-looking serifs which are uniform throughout the alphabet. The swash of the Q rests in simple grace below the perfect circle which comprises the main element and does not cut into it or across it.

ESSENTIAL CONSTRUCTION

If you have ever taken a course in mechanical drawing, you could render this style adequately even if you don't consider yourself the best letterer in the world. It is for this reason that Stymie Medium is such a favorite among engineers and draftsmen. Though this alphabet is suitable for brush and various types of lettering pens, its best characteristics emerge when it is done carefully, reproduction technique. This means ruling pen, compasses and straightedge are put to the task as well as precision and accuracy. In view of the fact that all strokes are of equal thickness, a template easily does the job of securing the desired uniformity. The work is carefully laid out in pencil and outlines of letters are ruled in with India ink. When that has been accomplished all that is left is a careful fill-in job with an appropriately sized brush.

ESSENTIAL APPLICATIONS

The alphabet shown on the opposite page is a standard type face and is to be found in the job case in most composing rooms. Its mechanical structure suggests stability, precision and masculinity. It is as appropriate when shaped out of wrought iron and stainless steel as it is when printed on book jackets, posters and brochures. Because it is a rugged letter that has inherent structural solidity, this type lends itself to the power saw machine. As for its adaptability for screen processes—Stymie Medium is a natural! With few exceptions the strokes are either straight or circular and can be made with straightedge and compass. One big reason why Stymie is one of silk screen's best friends is that it looks good (and prints well) with either direct or reverse printing. There are no fine lines to break away in adhering or to fill in during printing.

ABCDEFGHIJKLM
NOPQRSTUVWXYZ
&!?$1234567890

ESSENTIAL CHARACTERISTICS

Stripped of its deep shadow, Thorne Shaded is a Roman thick-and-thin letter with slab serifs. It has a unique character, other than the shadow, which gives it its name. While most serifs are thin slabs, some letters sport another kind of serif such as is evident in the C. This rather bulky triangular shape is repeated three times in the E, one for each cross stroke. It is also to be seen in the G, S, T and Z. Of note too, is the circle terminal of the J and the odd shaped swash of the Q.

As to the shadows, they are all consistently to the right and below the letter form, indicating a source of light at the upper left hand side of the letter.

Thorne Shaded is an open faced letter form, formed by a thin outline on one side and deep shadow on the other.

ESSENTIAL CONSTRUCTION

For hand lettering purposes, the outline would in most cases be disregarded and the letter treated as a solid. A practical reason for this modification is the fact that it is too time consuming to put a thin outline around each letter. If an outline should be desired, it can be done *after* the letter is blocked in a solid color.

To assure horizontal uniformity in the bottom depth of shadow, it is advisable to rule in a horizontal guide line. The side shadow can be measured off with a template or judged by eye for uniformity in thickness. You will note that the side shadow is thinner than that of the bottom. This is structurally necessary since to give the shadow its full depth would fill up the counters of the letters, creating blocky black areas.

For reproduction lettering, ruling pen and straightedge are recommended. The outline of the outer rim of the O and Q, being mechanically round, may be inked-in with a compass.

ESSENTIAL APPLICATIONS

Thorne Shaded was designed as a display face and should be restricted to headline copy, focal words in copy, but never to carry a complete selling message. In addition to its use in printing and reproduction, this is an excellent source alphabet for sign work on windows, trucks and displays. The open face strongly suggests the use of a color other than shadow color. For gloss work, gold leaf may be that other "color." Or if on glass may be treated as an outline letter exactly as shown here, but the inside areas within the outline could be very effectively treated in a translucent wash technique.

Thorne Shaded, because it is a difficult alphabet to letter by hand, is not suitable for showcards or hand painted posters, unless of course it is used with good judgment and confined to one or two "stand out" words.

ABCDEFGHIJKLMNOPQRSTUVWXYZ
abcdefghijklmnopqrstuvwxyz

ABCDEFGHIJKLMNOPQRSTUVWXYZ
abcdefghijklmnopqrstuvwxyz

&!?$1234567890

ESSENTIAL CHARACTERISTICS

Here's a type face rediscovered by contemporary designers and featured by the best composing rooms in the country. Torino, an import from Italy, is a strong, sturdy type face in spite of its rather classic Roman heritage. What are the chief characteristics which distinguish this from other Roman type faces? The hairlines and triangular serif terminals of the E, F, G, etc., the swash tail of the Q which does not cut into the main body of the letter itself, the unique formation of the G with the triangular serif on top and the interesting right side stroke with the tapered terminal. Also interestingly different is the center stroke of the W which cuts into the two V-shaped elements which comprise that letter.

The lower case alphabet reflects the same simple elegance and classic dignity of the capitals. The most unusual letter of the lower case is the g with its open lower loop turning gracefully into itself.

ESSENTIAL CONSTRUCTION

Torino is a purist's ideal in modern Roman alphabets. This imported Italian type face originated in a country famed for its classicism and expresses the many subtleties associated with that ancient heritage. The thin lines are hairline in weight, as are the fine serifs. Only the meticulous use of the ruling pen could produce the uniformity of the thin elements with the unvarying precision required. For the round elements such as the lobes of the B, the curves of the C, D, etc., no drafting instrument can be employed. To render these shapes well, a lot of talent, patience and an innate sense of design are necessary. As a type face, Torino is available from 8 to 48 points in both upper and lower case as well as in italics.

Torino Italic is shown here, by way of comparison to the standard Roman version. The upper case of the Italic bears a close family resemblance to the parent alphabet. The lower case however, departs so radically, as to constitute a distinctive type face all its own, retaining only the pristine elegance in the sharp definition of thin and thick strokes.

ESSENTIAL APPLICATIONS

When you are in a rush to get that hand-lettered job out of the house fast, Torino will not do. Torino refuses to be hurried! If duplicated by hand (time and price permitting) this beautiful type face will grace the display board of the finest exhibition panels. If intended for books and brochures, title pages and covers for annual reports and such typographic uses, Torino will be refreshingly different and most appropriate. This alphabet is not a practical one to use on hand-cut film stencils, but it reproduces well photographically. For that everything must be right—a good stencil, paint mixed just right, a smooth coated stock and loving care. Now that you have been introduced to this type face, put it aside and mark it reserved for that next Tiffany job that comes along.

ABCDEFGHIJKLMNOPQRSTUVWXYZ

&!?$1234567890

abcdefghijklmnopqrstuvwxyz

ESSENTIAL CHARACTERISTICS

For delicacy and grace, Trafton Script does not have many rivals among modern practical type scripts. The flowing lines suggest the quality of a pen. Trafton Script is based on a personal handwriting with lines that are thick or thin depending upon the intended accents of the strokes. This type face is too free to present a bill of particulars for each letter of the font. Each letter is distinctive and yet so varied, but all have this in common. There is abundance in the use of flourishes as in the A, B, D, H, etc., some of which extend beyond the confines of main guide lines. The angle is retained throughout, as is also, of course, the delicate contrast between the thick and thin elements. If just a few letters were to be singled out to best exemplify this type, it might perhaps be the A, B, D, R and Z.

ESSENTIAL CONSTRUCTION

Most typographers carry Trafton Script as one of their standard type faces in varied sizes in both upper and lower case. For purposes of hand lettering either for comp type indication or for color lettering over display panels, a fairly personalized version of this model will serve the task on hand adequately. In fact, for jobs definitely requiring hand lettering, it would be unnecessary, perhaps inadvisable to attempt a line-for-line, stroke-for-stroke copy of the specimen shown here. A more personalized and more spirited representation would be assured if the letterer were to analytically study the alphabet and then proceed to make his own, based on his own personal handwriting. To do this most effectively, a broad-nibbed pen or well-sharpened chisel-edged pencil would be the most yielding tools. This type face was not selected as a model for copy work but as a source of inspiration for individual creative effort.

ESSENTIAL APPLICATIONS

Informality and grace are the keynotes of this type face and it should be used wherever those effects are the twin objectives of the designer. A case in point may be the title page of a book on poetry or a heading of a story in a ladies' journal or in fact, any graphic appeal to the more delicate sex. The designer of this type face, Howard Trafton, would however, not want to see his creation so limited in scope, and a cursory examination of advertisements will indeed reveal the use of Trafton Script for all types of advertisements. It is also "at home" on letterheads, direct mail pieces, labels, chapter headings, for banks, etc. It has however no prominent place in poster or display advertising.

ABCDEFGHIJKLM
NOPQRSTUVWXYZ
1234567890
&!½?¿¡

ESSENTIAL CHARACTERISTICS

This type face, modelled after Playbill shown on page 171 represents an open face version skeleton letter formed by outline and shadow. As in Playbill, the heavy slab serifs give this type its special character. The upright strokes are comparatively thin, while the horizontal strokes (and serifs) are heavy, a reversal of the order of nearly all conventional lettering styles.

The letters are condensed and tall for their width, with all lobes flattened to rectangular shapes. The counters of most letters are uniform in width, thus each letter (with the exception of the I, M and W) takes up about the same horizontal space. Note too, that no letter stroke leaves the confined area of vertical space. It is customary in most letter styles to have some swash elements such as the R and Q extend beyond this area, but not so here. Note how the swash element of the Q has been tucked into the basic letter form so that it too does not break the line of vision.

ESSENTIAL CONSTRUCTION

In laying out this type face, it is often helpful to rule in not merely two guide lines (top and bottom), but four parallel lines. The two additional lines serve as guides for the thickness of the top and bottom block serifs. Shadows should be included *after* the layout is completely finished.

While the alphabet shown here is an "open letter" type face intended as a see through letter form, the letterer is free to modify it to the extent of using an additional color, as a fill-in between the skeleton outline. In such a case, it would be best to paint it in as a solid letter form and then add outline and shadow in a different color. In heightening the three-dimensional effect, the shadow may be modified in thickness, the heavier the shadow the greater the illusion of depth of letter. Also the position of the shadow may be altered so that it falls consistently in the lower, rather than the upper part or completely reversing it so that the shadow falls on the left.

ESSENTIAL APPLICATIONS

There are several type faces which resemble Trylon Shaded, shown here. One is a solid version, the others are Figaro and Barnum. All have a similar structural anatomy and may be used as alternates.

Of recent years there has been a strong revival of old faces, most of which represent modifications of type faces and hand lettering styles which were used in the days of the Old West. This typographic anachronism also has its counterpart in the use of time-honored engravings taken from old newspaper advertisements, illustrations taken from the pages of old, out of print books on anatomy, architecture, and old Sears and Roebuck catalogues.

The most obvious application of the type face shown here would be any merchandising which reflects a nostalgic past—the old time circus, western movies, T.V. titles and other relics of the days of the Golden West. To be more effective, Trylon Shaded should be used sparingly as a foil for a lettering style which is ultra modern.

ABCDEFGHIJKLMNOPQRSTUVWXYZ

&!?¢$1234567890

abcdefghijklmnopqrstuvwxyz

■ ESSENTIAL CHARACTERISTICS

This standard type face which is almost self-identifying, represents a blown-up version of the type used on office typewriters for years. Typewriter and its variations (Typewriter Remington, Typewriter Underwood, Typewriter Reproducing) are available in point sizes ranging from 8 to 48 points. Basically, Typewriter is a flooded, one-thickness letter with serif appendages. The slight quivering structure, the flooded serifs that flow into the stems, the unvarying dimensions of each letter regardless of its construction needs, all carry out the intended illusion—namely, to duplicate the highlights (and shortcomings) of typewriter type. Note the squeezed structure of the M and W, the expanded I and J and other such eccentricities of typewriter type which are now uncritically accepted as standard.

■ ESSENTIAL CONSTRUCTION

This is a simple letter to duplicate by brush or pen. When using a brush, be sure that it is fully charged with paint so that the terminals of the strokes come out full and flooded. For a pen version of Typewriter Bulletin, use a broad round or oval Speedball pen. Since it is desirable to retain the effect of typewriting, it would be helpful to do this. Typewrite your copy on a piece of paper first. Then you'll have a model to follow both for construction and the rather odd spacing which we are now accepting as standard for typewritten matter. These oddities which are unique to typewritten characters must be rendered exactly when reproduced.

■ ESSENTIAL APPLICATIONS

This type is a "natural" for any ad dealing with office procedure or equipment. It will awaken a particular professional and personal response in the feminine sex who are or have been secretaries or office workers. Also it is a good "gimmick" type to use for ads intended to reach the "big business executives."

The "Typewriter" type face has further application in advertising posters and promotional pieces to emphasize a particular message or thought in a business-like procedure. Used in this fashion, it can be combined with other lettering styles. A word or line of greatly enlarged Typewriter, shaky and fuzzy as it may appear, can serve as a foil for small sharp copy set up in a conventional type face such as Bodoni or Caslon.

ABCDEFGHIJKLMNOP
QRSTUVWXYZ
&!?$1234567890

ESSENTIAL CHARACTERISTICS

Umbra means shadow. This alphabet is one of our standard type faces and is self-identifiable because it is structurally composed of deep slanting shadows of a thin Gothic letter, similar to the Kabel type face. Squinting your eyes slightly will disclose a simple lettering style without serifs and free from frills or other embellishments. In a sense, the letter is latent within the shadow itself. The shadows of Umbra fall consistently to the lower and right-hand side of the latent letter structure.

ESSENTIAL CONSTRUCTION

It is well to bear in mind that Umbra is a traditional type face and as such is readily available in many forms—as a printer's type face, photo lettering style, cut letters in paper, etc. If, however, there appears to be a special need to render this style by hand on painted bulletin work and other direct application uses, the best suggestion is to first draw in a one-thickness Kabel type entirely devoid of shadows. This will give the basic structure of the lettering. When that is complete then add the shadows. The Umbra style lends itself very nicely to two-color work. In such a case the structure of the basic letter is rendered in one color and the shadow in another (usually the darker of the two colors).

ESSENTIAL APPLICATIONS

Umbra is included in this book, not because it represents a personal favorite of mine, but it would be remiss of me not to include it quite regardless of personal sentiments. In general it is a difficult letter to render by hand and a difficult alphabet to read. If Umbra is specified, its use should be limited to a line or two of copy in order to produce the desired effect.

An entire ad done in Umbra would produce a bewildering maze of shadows and confusion. With limits of propriety this style may have its place in titles of books, headings and wherever a three-dimensional effect is desired.

ABCDEFGHIJK
LMNOPQRSTU
VWXYZ
abcdefghijklmnopqrstuvwxyz
&!?1234567890$

ESSENTIAL CHARACTERISTICS

Here is one of the most popular type faces in use today. It is not likely that this alphabet will ever pass into oblivion, though its present popularity will perhaps diminish with the passing of time and the emergence of a new idol. Essentially it is a basic Gothic letter form. Although it appears at first glance to be a one-thickness letter, a more careful examination will reveal slight variations in thickness of stroke. This is easier to detect in such letters as the C, where the top and bottom curves thin down somewhat. Variations in thickness can also be noticed in the horizontals of the E, compared to the vertical. This is characteristic of most of the letters. The letters of this type face are greatly extended giving a compact black face appearance to a line, especially when it is reduced to the smaller sizes.

ESSENTIAL CONSTRUCTION

Most of the letters fit into a wide rectangle, although, as we have said before, they are unequal in stroke thickness. It is an easy letter to render by hand, but it is equally easy to get this alphabet in cold metal type or in some of the varied forms of photo lettering. Most composing rooms carry a plentiful stock of this type in point sizes ranging from 8 to 84 points. If rendered by hand for reproduction purposes, the mechanical nature of its construction suggests the use of ruling pen. No reliance on compasses is possible, since the round elements are not mechanically circular in shape.

ESSENTIAL APPLICATIONS

Because Venus is so massive and bold and demands attention, it is an excellent choice for posters and newspaper display ads. But it is equally at home on letterheads, business forms, etc. It is a fine letter to employ for all purposes where legibility is important. This includes display boards, lettering for trucks, exhibition panels, etc. Don't bury this alphabet in your dead files. Keep this sample handy for ready reference. It is a good addition to your growing list of service alphabets.

Venus Extrabold Extend'd

© 1977 por Ediciones del Castillo, S. A.
Esta edición ha sido preparada por
Ediciones del Castillo, S. A.
Marqués de Monteagudo, 16. Madrid-28
Depósito legal: M. 6.530-1977
Compoprint, S. A.
Marqués de Monteagudo, 16. Madrid-28